GENERAL LUCIAN K. TRUSCOTT JR

First published 2025

Exisle Publishing Pty Ltd
PO Box 864, Chatswood, NSW 2057, Australia
226 High Street, Dunedin, 9016, New Zealand
www.exislepublishing.com

Copyright © 2025 in text: Glyn Harper

Glyn Harper asserts the moral right to be identified as the author of this work.

Copyright in photographs as listed on page 191

All rights reserved. Except for short extracts for the purpose of review, no part of this book may be reproduced, stored in a retrieval system or transmitted in any form or by any means, whether electronic, mechanical, photocopying, recording or otherwise, without prior written permission from the publisher.

A CiP record for this book is available from the National Library of New Zealand.

ISBN 978-1-923011-01-4

Designed by Nick Turzynski, redinc. Book Design, www.redinc.co.nz

Typeset in Newzald Book 12/16

Printed in China

This book uses paper sourced under ISO 14001 guidelines from well-managed forests and other controlled sources.

10 9 8 7 6 5 4 3 2 1

While every effort has been made to ensure that the information in this book was correct at time of going to press, errors and omissions may occur. The author and publisher welcome any correspondence on such matters, but do not assume and hereby disclaim any liability to any party for any loss, damage, or disruption caused by such errors or omissions.

GREAT COMBAT COMMANDERS

'Quite a talent for fighting'

—— ¶ ——

GENERAL LUCIAN K. TRUSCOTT JR

GLYN HARPER

Contents

List of maps **5**

Map key **6**

A note on American, British and German army units and formations of the Second World War **7**

Prologue: an unwanted command **9**

1 Lucian Truscott's preparation for war **13**

2 Over there **35**

3 Commanding the dogface soldiers **48**

4 Disaster at Anzio **75**

5 Anzio command: Truscott's finest hour **92**

6 Fighting in France, August to October 1944 **119**

7 Back to Italy **138**

8 After the war **168**

9 Assessment **179**

Epilogue: Taps **186**

A note on sources for this book **188**

Further reading/watching:
Lieutenant General Lucian Truscott Jr **190**

Photographic credits **191**

Acknowledgements **194**

Index **196**

List of maps

Operation Torch: invasion of French North Africa **42**

The attack on Mehdia and Port Lyautey **44**

Operation Husky: the invasion of Sicily **55**

Sicily from Licata to Palermo **59**

The race to Messina **61**

The Italian battlefields **67**

Fifth Army on the Winter Line **71**

The Anzio Landing and ground captured **80**

The breakout from Anzio **106**

Operation Dragoon: the landing in southern France **125**

The pursuit up the Rhône Valley **128**

Fifth Army breaks out of the Apennine Mountains, April 1945 **161**

Map key

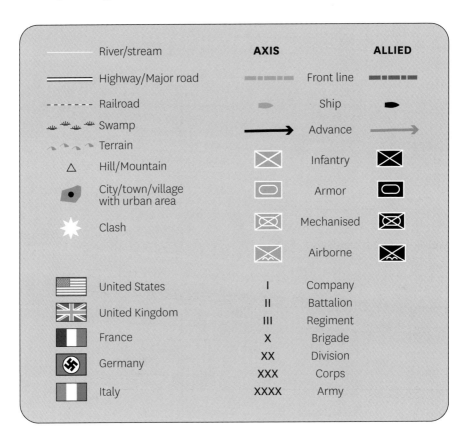

A note on American, British and German army units and formations of the Second World War

Technically, smaller army groupings up to battalion size are units or sub-units. The level above battalion size, which is a brigade or a regiment and higher, are technically known as formations. Many books do not differentiate between these classifications, simply referring to all army groupings as units.

The most important army fighting formation was the division. It was the number of divisions by which the size and strength of armies were judged. In the Second World War most divisions numbered around 15,000 men, although some could have more than this. Two or more divisions made up an army corps, usually denoted by the Roman numeral number of that corps. Two or more corps made up an army; two or more armies formed an army group.

German panzer divisions were an all-arms formation of tanks, motorized infantry and artillery. They were similar to the Allies' armoured divisions. The Germans also had panzer grenadier divisions which had fewer tanks and more infantry soldiers.

American and German divisions were divided into regiments while British divisions were divided into brigades. Most divisions contained three regiments or brigades. The division was usually equipped to operate independently once deployed by having its own artillery, engineers, medical, supply and transport units embedded in the division.

Regiments and brigades contained three core elements which, in the case of infantry, were called battalions. An infantry battalion was about 850 men strong and was itself divided into companies of about 120 men.

The Americans had the habit of adding additional armour and artillery to a regiment in order for it to achieve a specific task. These were then designated as a regimental combat team (RCT).

This explanation is needed as Lucian Truscott is the only American officer of the Second World War to command formations ranging from a regimental combat team to a field army.

Prologue: an unwanted command

Major General Lucian Truscott Junior was not happy that morning. It was 22 February 1944 and Truscott was serving as the deputy commander at the Anzio beachhead. He had been in the job for just five days. The Anzio beachhead was a narrow lodgement on Italy's western shore made by the Fifth Army's VI Corps in a daring amphibious landing that positioned them just 45 miles from Rome. But the landing, made on 22 January 1944, had not advanced far and it had been heavily resisted by the Germans once they had recovered from their initial shock. Already some senior commanders, including the man in charge at Anzio, Major General John Lucas, were comparing the Anzio landing to the ill-fated Gallipoli venture nearly 30 years previously. On 18 February Lucian Truscott had been replaced as commander of his beloved United States 3rd Infantry Division to serve as Lucas's deputy. It was a nothing job with no command authority and no clear role. Little wonder that he felt a bit of a spare wheel. Truscott was not feeling well either with a raging sore throat, not helped by too many cigarettes and little sleep. He was also suffering from a hacking cough, severe toothache and painful nasal polyps. But his ailments were not the main reason for his unhappiness. That morning Truscott was informed that the commander of Fifth Army, Lieutenant General Mark Clark, wanted to see him in the evening after Clark had first met with Lucas. Truscott guessed what this meeting would be about, and he was dreading it.

Lieutenant General Mark Clark met with Lucas at 2000 hours to inform him he was being relieved of command of VI Corps forthwith. Lucas was not surprised, expecting to be sacked ever since Truscott was appointed as his deputy. He wrote in his diary, with some bitterness, that: 'And I thought I was winning something of a victory.' To soften the blow, Clark stated that Lucas would be serving as his deputy army commander, in reality another nothing appointment, and that he would be leaving Anzio the next day. Clark then summoned Truscott.

At this new meeting Clark informed Truscott what had just happened and that he was now the commander of VI Corps and the Anzio beachhead. Truscott, never a yes-man, informed Clark that he had no desire to take over from Major General Lucas and believed the change was unnecessary now that the situation at Anzio had stabilized. Truscott also added that he thought sacking Lucas would have an adverse effect on morale and undermine the confidence of other officers. Clark overruled Truscott's objections, stating that the decision was final. Lucas would be leaving Anzio and Truscott was now the man in charge. Clark also confided to Truscott that he hoped to avoid wounding Lucas too much by the appointment to deputy commander. Truscott knew that any commander relieved on the battlefield would be 'deeply hurt'. His reply to Clark was blunt: 'You can't relieve a corps commander and not hurt him.' Truscott liked and respected Lucas who in turn had always treated him 'with the utmost consideration'. For Truscott this new appointment, even though it was a promotion to Corps command, was, in his words, 'one of my saddest experiences of the war'.

That evening in a villa at the town of Nettuno, after having his raging sore throat painted with silver nitrate, Truscott ate a late supper. He spent the rest of the evening in front of a fire with a bottle of Black & White scotch, a gift from Mark Clark to celebrate his new command. Truscott did not feel like celebrating. As he reflected on the day's events, Truscott felt both the burden and loneliness of a senior commander. He also had doubts whether he was up to the task, doubts he kept hidden except from one person. Some days later Truscott wrote to his wife Sarah about his new job:

> My new assignment came as a surprise of course — and at a most difficult time. During the past year I have of course become

greatly attached to the 3rd Division and of course hate to leave it. However, it is the fortunes of war, and after all, my one and only purpose is to serve my country. If this command offers a bigger opportunity, I must accept it even though I may feel my own inadequacy.

There was good reason for these feelings. Lucian Truscott knew his appointment as VI Corps commander at Anzio was something of a poison chalice and that his 'fortunes of war' could quickly be reversed. The tasks ahead of him were daunting. First, Truscott had to make the Anzio beachhead secure against future German counterattacks. To be driven into the sea would be a disaster and ruinous to the Italian campaign. Truscott had no doubt at all who would be blamed for such a humiliating defeat. Second, Truscott had to win back the trust of the many disgruntled British formations in VI Corps, especially those who had pushed hard for Lucas's relief. Third, he had to make sure he and the command team were visible to every unit on the beachhead. Finally, when the time came, Truscott knew that it would be up to him to plan, organize and execute the breakout from Anzio if the Allies wanted to avoid another Gallipoli-style fiasco. If he failed to achieve even one of these tasks, Truscott knew full well that he would suffer the same fate as Major General John Lucas.

The distinguished historian Barbara Tuchman wrote of the critical importance of combat commanders and the challenges they face:

> When the moment of live ammunition approaches, the moment to which all his professional training has been directed, when the lives of men under him, the issue of the combat, even the fate of a campaign may depend upon his decision at a given moment, what happens inside the heart and vitals of a commander? Some are made bold by the moment, some irresolute, some carefully judicious, some paralyzed and powerless to act.

For combat commanders, these are the very fortunes of war. Major General Lucian Truscott Junior's place as a senior Allied commander and the fate of the Anzio beachhead depended on how well he performed in the weeks ahead.

1 Lucian Truscott's preparation for war

Early days

Lucian King Truscott Junior was born on 9 January 1895 in Chatfield, Texas. Truscott's grandfather had been an immigrant from Cornwall, England, while his mother Maria's family traced their origins to Ireland. Although Truscott's family moved to Oklahoma six years after his birth, Lucian Truscott Junior always claimed to be a Texan.

While living in Chatfield, the young Lucian had an accident that nearly killed him and that marked him for the rest of his life. Lucian's father, Lucian King Truscott Senior, was the town's physician. One day, while Lucian Senior was busy in another room, the young boy decided to play in his father's office. Spying something that looked good to drink, Lucian swallowed some carbolic acid. The child's scream brought his father running. Dr Truscott managed to save his son's life, but the accident left Lucian with a raspy, gruff voice for the rest of his life. It would later become one of his trademark features, with one observer claiming that Lucian Truscott Junior had a voice like 'a rock-crusher that gives his orders an awesome ferocity'.

The Truscott family moved to Oklahoma in 1901. They were not well off and kept moving from town to town in frontier Oklahoma every three or four years. The reason for this constant relocation was, until recently, a closely guarded family secret. Historians speculated about Lucian Senior's continual need to move on and suspected he had accumulated gambling debts or had a string of failed business ventures. Lucian Senior had certainly speculated on racehorses and farm property, all of which ended in disaster, but there were more serious concerns. The reality was that Lucian Senior, the country doctor, was a heavy drinker and had also become addicted to laudanum. Laudanum, or tincture of opium as it was known, was a mixture of alcohol and opium and was in common use in the United States until well into the twentieth century. It was principally used to treat pain and as a cough suppressant. Once Lucian Senior's drinking and drug use became widely known, the family needed to move on.

This constant relocation was particularly hard on young Lucian's mother, Maria, a college graduate who was forced by circumstances to home school the children. To help support the family, which now included three sisters, young Lucian, a sister and his mother worked chopping and picking cotton. Then in 1911 both Lucian and his mother attended Summer Normal School at Norman, Oklahoma, in order to acquire a teaching certificate. The 'normal' schools later became teachers' colleges. Lucian Junior was successful and at the age of sixteen, having advanced his age by two years, he was teaching in local schools, first at Stella and then later at Onapa and Eufaula. For six years Truscott taught in schools in several small towns in Oklahoma state. One historian has described this period of teaching in one-room rural schools as 'one of the hardest apprenticeships imaginable'. But what was the young Lucian Junior like? His sister Patsy described him at Eufaula as being quite religious, helping build the First Christian church there and insisting that its doors always remain open. She described her brother as 'just a normal healthy specimen of American youth, only different in that he did not drink, smoke, nor swear ... [and] he hadn't much time for play.'

Joining the army

Lucian Junior, however, wanted a more active career than teaching offered. His dream had been to attend the West Point Military Academy, but his family could not afford the fees and a scholarship was unlikely. Instead, service in the Army Reserve programme offered an alternative route as well as a taste for military life. In 1917, after two years' service as a lieutenant in the Army Reserve, Lucian Truscott Junior, aged 22, applied to become a Regular Army officer. In support of his application Truscott submitted several letters of recommendation. The one from the superintendent of the McIntosh County Schools was typical, describing Truscott as 'a young man of honor, integrity and capability'. Truscott's timing was fortuitous. The United States was experiencing border problems with Mexico, including the attack on Columbus, New Mexico, by Pancho Villa in March 1916, and in April 1917 it had declared war on Germany. There was now a pressing need to expand the US Army, including its officer corps. The National Defense Act of 1916 created several new programmes to enlist men and to recruit officers from outside the usual channels of military academies. Lucian Truscott Junior applied to attend the first of these officers' training camps of the First World War which would become the forerunner of the officer candidate schools (OCS). Truscott easily passed the written and physical examinations and subsequently received his orders to attend a basic officer course at Fort Logan H. Roots in North Little Rock, Arkansas.

The course at Fort Logan H. Roots was three months long and it was intense. As one prospective officer recorded of it: 'Everything is business, and there is not time to fool around. It is go from 5:30 in the morning to 10:30 at night, with very little sleep.' There was a reason for the frantic pace of this course. It aimed to produce an army junior officer ready to command men in just 90 days, something that it took West Point four years to achieve. Lucian Truscott Junior graduated top of his class as one of the 'ninety-day wonders' as they were called and was offered a provisional commission as a second lieutenant. Even though he had topped the course, Truscott did not feel he had been adequately prepared. He later wrote: 'Even after our training, our military background was sparse. Most of us were completely ignorant of things military and had never seen an organized unit of the army or a Regular Army officer until admission to the training camp.'

The branch of the US military Truscott chose was the cavalry, then considered a small but elite part of the army. The National Defense Act had authorized the establishment of ten new cavalry regiments. It was a life-changing moment for Truscott. As he confessed to a *Life* magazine reporter almost 30 years later, joining the Army in 1917 had 'rescued' him from a life of obscurity. The new officers' training camp had provided Truscott with a backdoor entry to a commission in the United States Army, albeit a 'provisional' one. It was up to him to make the most of this golden opportunity.

Truscott's military file noted that when he joined the army, he was five feet, nine inches tall and weighed 170 pounds (175 cm; 77 kg). But he seemed much bigger than that and carried himself with a considerable degree of authority. His son, Lucian Truscott III, has left a vivid description of his father:

> He was a handsome man, attractive to women, but not big, being perhaps five foot-ten and about one hundred and eighty pounds when he was in good physical condition. But he *seemed* like a big man. He had large eyes, a prominent nose, large but not protruding ears, broad shoulders, a big chest, and huge hands with big, square fingers.

Second Lieutenant Lucian Truscott Junior did not see active service in the First World War. Instead, he served with 17th Cavalry Regiment on the border with Mexico near Douglas, Arizona. The 17th Cavalry was stationed at Camp Harry J. Jones, a sprawling 'sea of canvas' stretching for more than two miles from the town of Douglas to the high barbed-wire fence that marked the Mexican border. Truscott later recalled that two smells dominated the town of Douglas: acrid sulphur from two great copper smelters west of the city and the odour of 'stables, horses and leather' so familiar to cavalrymen.

Truscott and the other newly minted 'ninety-day wonders' soon discovered they had much to learn. Their military education at the training camp, Truscott admitted, had been 'austere and elementary', conducted by instructors 'who seemed to know little more than the candidates'. They had much to learn from the professional soldiers at Camp Jones. This included the correct military way to ride a horse, the daily routine of morning reports, duty rosters, sick parades and unit administration. It also included using the correct terminology and Truscott recalled earning a harsh rebuke from his stern commanding officer for referring to the soldiers under his command as 'boys'. 'Mr Truscott, they're men, damn it!' the Colonel yelled. 'They're men! Every one of them! They're men! Men! MEN!' Truscott did not need to be told twice.

Lieutenant Colonel James H. Hornbrook, Truscott's first commanding officer, was, as Truscott later recalled, 'a ruthless disciplinarian, he was strict, abrupt, and treated words as they were drops of water in a canteen in the desert'. Hornbrook may have been a man of few words and short on empathy, but he taught young officers like Truscott invaluable lessons. These included how to analyze and solve problems, the importance of relying on a troop's first sergeant for knowledge and advice, teamwork and even how to load pack animals correctly.

Life at Camp Jones was regulated by bugle calls which had to be obeyed. As Truscott recalled, the bugler 'ruled our lives with the clear notes which penetrated every corner of the camp'. The first bugle call was Reveille which sounded at 0530 hours. Then followed other calls throughout the day to signal Mess (mealtimes), Police Parade, Sick Parade, Drill, Recall, Stables, Officers and First Sergeants with the final two calls of the day being Retreat in the evening and Taps at 2300 hours. All in camp followed the bugle's orders.

Truscott was disappointed not to serve in France and worried that when the war ended, he would be forced to return to teaching. However, on 20 March 1919, 17th Cavalry Regiment received orders to move from Douglas, Arizona, to San Francisco. From San Francisco the regiment was to embark on an army transport in early April to sail to Hawaii on garrison duties. Truscott would be going with them. This was the start of what his son Lucian III described as 'normal cavalry assignments during the first two decades of his career'.

Marriage

Travelling with Truscott to Hawaii was his new bride. Truscott had met Sarah Nicholas Randolph through her father, an army doctor who had treated the young officer for pneumonia. When Lieutenant Lucian Truscott had recovered, Dr William Mann Randolph invited him to his house for a family dinner. His daughter, Sarah, was staying with him at the time while on a break from her college studies in Georgia. Truscott and Sarah came from completely different worlds. Sarah was a fourth-generation granddaughter of one of the founding fathers of the American republic and who later served as its third president: Thomas Jefferson. She had been raised at Edgehill, a plantation four miles east of Charlottesville, and within sight of Thomas Jefferson's stately Monticello mansion. Sarah's family was well-connected and wealthy; one of the most prominent families of Virginia. Her son Lucian Truscott III wrote that his mother was a true Southern belle similar to Scarlett O'Hara of *Gone With the Wind*. She spoke with a soft southern accent with broad English A's that was unique to Charlottesville. No wonder that Lucian Truscott Junior was attracted to Sarah.

In 1919 Sarah Randolph was 23 years old and has been described as 'vivacious and beautiful'. She had many potential, well-connected suitors yet she chose Truscott, a lowly second lieutenant who was the son of a disgraced country doctor. Yet Truscott also had his attractions. As a young army officer, he was strikingly good looking. He was also intelligent, modest, sincere and clearly smitten by Sarah. Her son Lucian III wrote that Sarah 'had the fire, the determination, the will and ambition to convince my father he could succeed in anything he attempted'. It would be a true partnership and Sarah's role in it was crucial. As Lucian III wrote: 'She helped him become a gentleman who in later years would feel perfectly at ease with both infantry privates in foxholes and royalty in palaces.' It is revealing that in his wartime letters, Truscott invariably referred to Sarah as 'Beloved Wife' and drew strength from her faith in him. A letter written just before he was about to lead men into combat for the first time stated: 'My greatest ambition is to justify your confidence and to deserve your love.' Another written just before the landing on Sicily is typical: 'Your calm confidence in me is always with me and when doubt falls upon me — as it must on all — that thought soon restores that confidence. I can only do the best I can.' West Point graduate and successful novelist Lucian K. Truscott IV believed that it was Sarah who made her husband the general he would later become:

> My grandfather married the right woman. He was a redneck from nowhere. Without my grandmother at his side, he would never have been a commanding general at Anzio. He might have been a captain of the polo team, but not a general. With their Jefferson and Randolph background, she refined him so that he could one day be a general.

There were some early tensions though. Sarah's family were not happy that she was going to be an army wife and marrying a lowly second lieutenant without prospects. Sarah felt this pressure and admitted that she had planned to persuade her new husband to leave the army for something better. She later admitted:

> When I was married I was determined to try and get Lucian to leave the Army. My family did not feel that it was a profession with a great deal of future. . . . I fell in love with the life. . . . It would have been a sacrilege if I could have so influenced Lucian, because being an Army officer had been his dream since he was a tiny boy.

Lucian King Truscott Junior and Sarah Nicholas Randolph were married on 5 April 1919 in Cochise County, Arizona, just a week before their departure to Hawaii. Immediately after the wedding, Truscott was promoted to first lieutenant.

Sarah Truscott's influence on her husband and his career was profound. According to their son Lucian III, Sarah had just three interests in life. He lists them as 'her husband, her house, with its beautiful furnishings, and her children' then adds the comment that: 'without disparaging her priorities, I think her interests were in that order'. Lucian III could recall to the end of his life his mother sitting quietly at a crowded dining room table watching her husband become more and more boisterous, presumably as the alcohol flowed. During a pause in the conversation Sarah leaned close to her husband and said quietly: 'Lucian, you're *not* being very attractive.' As Lucian III recalled, his father's response was instant: 'Everyone would chuckle, the conversation would resume, and invariably he would be more subdued than before.' With her husband as the main priority in her life, Sarah did everything she could to assist his army career and kept him in line to do well. It is telling that when Lucian Truscott published his book *Command Missions* in 1954, he presented a copy to Sarah and had inscribed on the flyleaf: 'You were a far better soldier than I was over the years.'

On 19 July 1919, Truscott's status in the army improved considerably. After passing a final oral examination in eleven subjects, Truscott's provisional commission ended and he was appointed as a regular army officer. With a young wife and about to start a family, this was a welcome change for him.

Playing polo

In the early 1920s while stationed in Hawaii Truscott took up polo. He had never seen a polo game before coming to Hawaii where it was very popular. It soon became a lifetime passion. He was a skilled, aggressive,

The dust flies in a polo match between two United States Field Artillery units on 28 May 1926. The rider about to strike the ball and score a goal is Lieutenant Sharp of the 16th Field Regiment. This was the opening match of the War Department Polo Association Tournament. Truscott learned to play polo on a posting to Hawaii where the game was very popular. He was a skilled, aggressive polo player, being good enough to make the US Army team in international competitions. Playing polo was also helpful to Truscott's army career.

competitive player who showed complete disregard for his own safety. By the 1930s Truscott's prowess at polo had earned him a four-goal handicap and a place on the army polo team. In every polo match he played Truscott was determined to win and showed a ruthlessness that surprised many. Explaining to his fourteen-year son Lucian III why he had deliberately fouled an opponent to prevent him from scoring, Truscott stated:

Listen Son goddamit. Let me tell you something, and don't ever forget it. You play games to *win*, not lose. And you fight wars to win! That's spelled W.I.N! And every good player in a game and every good commander in a war, and I mean really *good* player or *good* commander, every damn one of them has to have some sonofabitch in him. If he doesn't, he isn't a good player or commander. And he never *will* be a good commander. Polo games and wars aren't won by gentlemen. They're won by men who can be first-class sonsofbitches when they have to be. It's as simple as that. No sonofabitch, no commander.

Playing polo was important to Truscott's army career, although it was not why he played the game. It gave him prominence in an army whose officer corps was dominated by West Pointers. Also, it gave him the opportunity to mix with these West Pointers on equal terms. One polo-playing friend with a style similar to Truscott's was George S. Patton. Patton, who had attended the Virginia Military Institute before moving to West Point, was senior to Truscott by ten years. Truscott would serve under Patton in Sicily in 1943 and then replace him in Germany in 1945. Polo was also partly responsible for Truscott's first senior appointment during the Second World War.

Army postings

The Truscotts spent just over a year in Hawaii during which two of their children were born. The US Congress passed the National Defense Act in 1920 which drastically cut their armed forces. While the ceiling for Regular numbers was 14,000 officers and 365,000 enlisted men, Congress approved funding for just 150,000, making the US Army the eighteenth largest in size and smaller than that of Belgium's. This meant that many army units were made inactive including three cavalry regiments with the highest numbers, one of which was the 17th Regiment in Hawaii. Lucian Truscott, newly promoted to captain, was posted to the 11th Cavalry Regiment stationed at the Presidio in Monterey, California. Truscott retained the rank of captain for the next fifteen years. As he later wrote, it was with a 'sense of sadness and regret' that they departed from Hawaii.

Other postings soon followed. From the Presidio, the Truscotts moved to Douglas, Arizona, then to Camp Marfa, Texas, where they were joined by Truscott's widowed mother. Maria Truscott lived for another sixteen years and spent most of that time with her son's family. From Camp Marfa the family moved to Fort Riley so that Truscott could attend the Troop Officers' Course at the Cavalry School. This was the start of Lucian Truscott's formal military education, and it was crucial for an officer's career that they do well at these army schools. Failure to graduate was terminal to a career. Only those who did well were candidates for further advancement. Exceptional students were invited back as instructors.

Truscott was one of 56 members of the Officers' Class of 1925 at the Cavalry School. Most were captains. As they arrived, the class members were encouraged to wear a distinctive new uniform consisting of a dark olive blouse, Sam Browne belt, knee-high boots and tan riding pants that flared at the waist down to the knees then tapered to fit into the riding boots. Truscott readily adapted to the new uniform and, when combined with a russet leather jacket and a paratrooper's white silk escape map knotted around his neck as a scarf, it became his standard dress during the Second World War.

Like most US Army education courses, the timetable was demanding with every minute of the day allocated to an activity. Class instruction lasted eight hours a day, but this was followed by several hours of study each night. All told, Truscott underwent 1400 hours of instruction in topics ranging from tactics and military law to weapons and horsemanship.

Truscott graduated from the Cavalry School in 1927, receiving his diploma with 'a thrill of pride in being a cavalryman and in knowing that he was a better one for having "been to Riley".' It was fortuitous that Truscott appreciated Fort Riley as he was immediately posted there as an instructor. Unlike other postings, this one lasted for six years. It must have been a godsend for Sarah after so many previous shifts and it enabled the Truscott children to have a decent stint of uninterrupted schooling.

Truscott the father

The Truscotts had three children and Lucian Truscott Junior was a rather stern, demanding and strict father. His son Lucian wrote that his mother Sarah 'was a loving person to whom touching us was as important as a vocal expression of affection'. Not so his father. Physical contact and any other expression of love for his children according to Lucian 'was rare'. When the children were young, Lucian Truscott did not hesitate to administer physical punishment with a razor strop. When they became too old for spankings, the usual form of punishment was being confined to their room. Lucian III recalled that his father insisted that rules were strictly adhered to. Told to be home by 10 p.m. one evening, young Lucian III arrived home at 10.02. He was immediately confined to his room for a week after school hours, but the confinement also included an entire weekend. Even though young Lucian's watch showed that he had a minute to go, Truscott's watch showed that his son was two minutes late. He had broken the rules and needed to be punished. During the confinement though, Lucian III was permitted to take a horse from the stables and ride so long as he did so alone. It was a tough lesson.

There is a strong hint in a letter written to Sarah, his 'Beloved Wife', on 14 May 1944 that Truscott had some trouble relating to young people, especially in their teenage years. In the letter Truscott reflected that it was exactly 27 years ago that he joined the army and that 'much water had flowed under many bridges since that May day'. Truscott told Sarah that he was feeling younger now than when he enlisted and then he admitted:

I'm sure that I have a much sounder appreciation of 'young people' psychology than I did then. Perhaps it would be strange if I did not but nevertheless one sees so many who do not maintain their contacts with youth.

It may have been this last point about 'maintaining contact' that saw Truscott writing frequently to his younger son Jamie. His letters were newsy and always encouraged Jamie to do well at school. He signed them: 'Your Devoted Dad'.

Posting to Fort Myer

The interwar period was a difficult time to be a horse soldier. It was clear to many that traditional cavalry would have a limited role to play, if any at all, in a future war. In 1926, while Truscott was attending Cavalry School, the War Department decided to reduce the cavalry branch numbers by 50 per cent. 'These were troublesome times for the cavalry branch of the army,' Truscott later reflected. While a student on the officers' course, the Chief of Cavalry, Major General Herbert B. Crosby, had warned the officers that:

> There is no use beating about the bush, because there is no doubt that the cavalry is on the defensive at the present time. . . . We are fighting for our lives and we have to keep fighting all the time.

For General Herbert the answer was to embrace mechanization which he believed would revitalize the role of the cavalry. Truscott was in full agreement, but many other officers felt that nothing could replace the mounted horse soldier. The cavalry's saving grace came in 1930 when the army's new chief of staff, General Douglas MacArthur, made the decision that developments in mechanized forces — that is, armoured vehicles — would be a cavalry responsibility. Soon after this announcement the oldest and youngest cavalry regiments, the First and Thirteenth, turned in their horses, moved to Fort Knox, Kentucky, and formed the First Cavalry Brigade (Mechanized). Truscott appreciated that the cavalry's future was in mechanization, but he felt some regret that the age of the horse soldier was passing.

In June 1931 Truscott was posted to Fort Myer in Arlington, Virginia. He was appointed as the commanding officer of Troop E of the Third Cavalry Regiment. Fort Myer was a small post primarily of cavalry soldiers. It was also the most visible military unit in the nation, being used as ceremonial soldiers for many state occasions. These included marching for presidential inaugurations, escorting foreign dignitaries and, most importantly, providing honour contingents for funerals at Arlington National Cemetery. With almost half of the families of deceased US servicemen in the First World War opting to bring their remains for burial in the USA, this meant the US Army had to conduct several funerals a day at Arlington. Duties also included the perpetual guard mounted at the Great War's Tomb of the Unknown Soldier. The Fort Myer posting was also a return to Sarah's ancestral roots and their third child was born in December. In early 1932, another officer was posted to the Third Cavalry Regiment as its executive officer. This was Major George S. Patton, fresh from his graduation from the Army War College. Also at Fort Myer in 1932 was the Chief of Staff, General Douglas MacArthur, and his assistant Major Dwight D. Eisenhower.

All four men would be involved in a rather shameful episode in the US Army's history. In May 1932, during the Great Depression, more than 20,000 angry Americans, most of them veterans of the First World War, and their families, had converged on Washington DC to demand early payment of their promised war bonus. While the World War Veterans Act, or 'Bonus Bill' as it was more popularly known, had authorized additional payments to First World War veterans, most would have to wait until 1945 to collect their payment. These men and their families were in a distressed state and refused to wait that long for money owed to them which was desperately needed. At Washington

Captain Lucian K. Truscott Junior in a photograph taken about 1931. The badges on the collar lapel are those of the United States cavalry.

DC they set up campgrounds in the military style and named their gathering the Bonus Expeditionary Force. Most people referred to them as the Bonus Army. The Bonus Army later occupied an abandoned building on Pennsylvania Avenue. They received little sympathy from the government and on 28 July President Herbert Hoover authorized the use of troops from Fort Myer to clear the Bonus Army. This was only the second time since the Civil War that US soldiers were used against its citizens. The force used was excessive and included a mounted cavalry troop from Fort Myer, an infantry battalion from Fort Washington and a few First World War tanks.

MacArthur, against the advice of Eisenhower, turned up to watch the show. The Bonus Army was soon in full retreat, driven out by tear gas, fixed bayonets and mounted cavalry with sabres drawn, and their camp sites set on fire. Truscott showed little sympathy for the hard-up veterans, writing of the burning camp that:

> It was soon a mass of flames. By morning, the veterans were gone, and the huge primitive camp was a smouldering mass. No one knew who set off the first fires, but it was the complete answer from both a sanitary and disciplinary point of view.

This was Truscott's first military action and he later reflected that: 'Cavalry training and special training for riot duty had paid off. The unruly mob had been dispersed without bloodshed, without animosity, and with comparatively little trouble.' It was not much of a victory and Truscott was wrong that there was no animosity. Many Americans, including President Herbert Hoover, were angered at the treatment meted out to the veterans. As Harvey Ferguson has written of this routing of the Bonus Army: 'the U.S. Army exuded the cockiness of the bully who has just run the new kid out of town'. It was an apt description.

The next time that Truscott rode a horse along Pennsylvania Avenue was a much happier occasion. On 4 March 1933 he was part of the inauguration parade for the newly elected president Franklin D. Roosevelt. It was the last time that a presidential inauguration included a mounted cavalry squadron as a presidential escort. At the beginning of 1934 Truscott was delighted to learn that he had been selected to play mallet number three in the US Army Polo Team in a forthcoming international tournament in Mexico City. A few months later came even better news. He was being posted from Fort Myer to attend the most important military education course for US Army officers, the Command and General Staff School at Fort Leavenworth, Kansas. He would be promoted to the rank of major upon arrival there. All officers attending the school knew that a good performance here was an essential stepping-stone to higher command. Truscott appreciated its importance, later describing the Command and General Staff School at Fort Leavenworth as 'perhaps the most important military institution in the United States Army'.

The Command and General Staff School

Truscott was part of the class of 1935–36 which had 109 students. The course, the equivalent of a postgraduate degree, was probably the hardest army educational programme at the time. So much was riding on how well a student there did. Truscott later recalled that: 'The course was intensely competitive. Class standings were posted regularly. Officers were made to feel that their entire future careers depended on their class standings.' The curriculum covered the essentials of command, leadership and staff procedures and had been broadened to include technical developments in motor transport, aviation, armour and mechanization. The days were long with instruction starting at 8 a.m. and finishing at 5 p.m. daily. Students were also required to put in two or more hours of intensive study each evening in preparation for the next day's topics. On Friday afternoons students were tested with a map exercise and their grades posted on

Monday mornings for all to see. Most of the students on Truscott's course did well. It was a class on which the stars that mark the rank of a general officer shone. Of these 109 officers, 52 would later reach the one-star rank of brigadier general or higher with three becoming three-star lieutenant generals.

The importance of a military education provided by US Army schools and especially the Command and General Staff School should not be underestimated. They forced army officers to think outside of their narrow branches and small unit tactics and to address how the army should develop to meet future threats. This was crucial to the transformation of the US Army that occurred in the Second World War. Russell F. Weigley has written of this staggering sea change:

> The American army's capacity to transform itself during the next few years was as impressive an achievement as any in military history. The achievement was possible in large part because the immense material resources of the United States were available to support it. It was possible also because the 12,000–13,000 officers of the old army had succeeded in preparing themselves mentally for the transition to a greater extent than the observer of mounted parades and manoeuvres — and polo matches — might have suspected. The officers did so thanks largely to an excellent military school system modelled on European examples and long embedded, somewhat incongruously, within the frontier constabulary.

Truscott did well at the Command and General Staff School. His personal situation improved too with a promotion to major halfway through the course. In addition to the promotion came a long-awaited pay rise for army officers which restored their pay to what it had been in 1928, prior to being forced to take a 15 per cent pay cut.

After completing his course in 1936 Truscott remained at the school as an instructor for the next four years. Only the very best students were chosen for this posting, and it marked Truscott as destined for promotion to at least colonel rank. While at Leavenworth and thinking that another war in Europe was likely, Truscott, with the aid of Sarah, began learning French. He reached a degree of proficiency to enable him to translate the *French Cavalry Journal* for the Command and General Staff School library. He also found time to help prepare Lucian III for the entrance examination of the US Military Academy at West Point.

Other interests

There was a softer side to Lucian Truscott Junior than a rising army officer and fearless polo player. He believed strongly in self-education and was always reading. He loved gardening and fresh flowers. During his campaigns in Europe during the Second World War, Truscott always tried to have a bunch of cut flowers on his desk. He enjoyed cooking as well as working with wood. He became adept at making beautiful furniture using only hand tools, and even made his own polo mallets. Lucian Truscott also taught his sons and later his grandchildren how to work with wood using hand tools. His son Lucian III said that: 'He had great patience and was a good teacher, of both small boys and of men.' He recalled his father used the same instructions with both him and his son: 'Look at what you're hammering Son, not the hammer or your finger.' Lucian Truscott also taught his sons how to play polo with the same dedication and patience.

In August 1940, Truscott left the Command and General Staff School with the strong feeling that the US would soon be involved in another European war. After receiving some advice from George Patton, but against the wishes of the then head of the chief of cavalry, Truscott opted to join the new armoured and mechanized force being developed. He was promoted to lieutenant colonel and posted to Fort Knox, Kentucky, as the executive and operations officer of the 2nd Battalion, 13th Armored Regiment, 1st Armored Division.

Working with Eisenhower on war games

With the outbreak of the Second World War in 1939 and the likelihood of American involvement, the US Army was undergoing rapid expansion, which meant accelerated promotion for regular force officers. Truscott had served as a captain for fifteen years but held the rank of major for just on five years. The rapid expansion at Fort Knox meant there were more officers and families than the available housing. For many months the Truscott family lived in a large army tent that had water available, but they had to share a communal bathroom. With no kitchen available, all their meals were taken at the officers' mess. With the approaching winter, Truscott was forced to rent an apartment in a nearby town.

Truscott served at Fort Knox for less than a year. He found it an invaluable posting despite its short duration, later writing:

> But even a few months of experience in the armored force brought knowledge of the capabilities and limitations of armor and some concept of the views regarding missions and methods of tactical employment. These months were therefore of the utmost professional value to the officers who had the experience.

In March 1941 Truscott was posted to Fort Lewis, Washington, to work with the commander's chief of staff, Dwight D. Eisenhower, on a large-scale war game that would be played out in California. The preparation and conduct of these exercises required a great deal of planning and staff work of which Eisenhower, who had topped his course at the Command and General Staff School, was a master. Truscott described the California war game exercise as unforgettable, adding that: 'They were so realistic that only the shot and sound of battle was missing.' After the war game was finished Eisenhower departed Fort Lewis to organize another larger war games exercise in Louisiana. This would be the largest military exercise ever held in the USA during peacetime. It took place over much of the state of

Louisiana as well as a large part of Arkansas. Truscott also took part in the Louisiana manoeuvres as an umpire. The war game revealed that there was an urgent necessity to upgrade the US armoured forces and find suitable commanders who knew how to use a mechanized force. Truscott concluded that the war games:

> were of the utmost value to those who participated and were of value to the high command in sorting out some of the commanders who were engaged in it. The enormous value of the exercises would be demonstrated in little more than a year.

Call to arms

Truscott was at Fort Lewis, Washington, for just eight months. Orders came posting him back to Fort Bliss, Texas, for duty as commanding officer of the 5th Regiment, 1st Cavalry Division. Truscott found the post little changed since he had served there almost seventeen years earlier. Two aspects that had changed dramatically were that the motor vehicle had replaced horses and the division possessed considerable firepower. The 1st Cavalry Division was equipped with 17 combat cars (tanks), 178 scout cars, 180 trucks, 420 motorcycles, anti-tank guns, mortars and hundreds of light and heavy machine guns. Truscott was stationed at Fort Bliss for just on a month when, on a sleepy Sunday in early December, news came that the Japanese had attacked the American fleet at Pearl Harbor. Truscott recalled that the news was received with disbelief, then shock 'that finally gave way to one of bitter anger'.

Seventeen days after the Pearl Harbor attack Truscott was promoted to full colonel then things went quiet. He would not hear news of any further appointments for three months. Then on 1 April 1942 came a call from the War Department in Washington DC. The man on the phone was Brigadier General Mark Clark who asked Truscott how soon he could leave to take up an important assignment. Truscott replied that he could leave straight away. Clark replied that he had three days to get to Washington and that he should prepare for an extended overseas posting in a cold climate. Not even in his wildest dreams would Colonel Lucian Truscott Junior have envisioned the assignment that the War Department and the Chief of Staff of the United States Army, General George Marshall, had in store for him.

2 Over there

Reporting for duty

When Truscott reported to Mark Clark at the War Department in Washington DC he was startled when Clark asked him how he would like to be a British commando. Truscott, like most American soldiers, knew little about British commandos apart from the fact that they received rigorous training and were regarded as elite troops. He later recalled: 'In spite of my astonishment, I was able to say that I thought I would like it very much.' Clark then explained that Truscott had been chosen to lead a small team of American officers who would be attached to the British Combined Operations Headquarters then commanded by Lord Louis Mountbatten. Their tasks were to observe the organizational and training structures of the British commandos with the view to set up a similar unit within the US Army. They were also to observe the preparations for an amphibious landing being planned by Mountbatten's headquarters and to try to expose as many US officers to combat as was possible. It was a plum assignment and, as Truscott wrote in *Command Missions*: 'I could hardly have been more amazed.' Clark informed Truscott that he would receive further briefings from both Eisenhower and General George Marshall.

At the meeting with Eisenhower, who was heading the War Plans Division of the General Staff, Truscott learned that it was Eisenhower who had recommended him for this new assignment. When he heard that part of his new mission was to study amphibious operations, Truscott expressed his reservations. He was a cavalry officer who had been on a small boat in open water just twice in his life. Eisenhower explained that Truscott was as well fitted as any American officer to take on this role, especially as he had been an instructor at Fort Leavenworth. But it was Truscott's skill as a polo player that had singled him out. 'Did you know that Lord Louis wrote a book on polo? You can learn, can't you?' Eisenhower asked. The decision was made. Truscott was heading to the United Kingdom.

Before he left the United States, Truscott read all he could about British commandos, planning amphibious operations and British military organizations. He also had a meeting with General George Marshall just before he departed. Chief of Staff George Marshall had many skills, one of which was his ability to identify men of talent who would be suitable for high command in war. He had singled out Eisenhower, Patton and Clark but not Lucian Truscott Junior. Marshall, who never once smiled during the interview, alarmed Truscott by opening with the statement: 'You are an older man than I wanted for this assignment.' Truscott remained silent. Marshall continued: 'Some of your friends are mad at you, nevertheless you have the characteristics I am looking for.' Marshall then outlined that he wanted Truscott to absorb all he could about planning amphibious operations. He informed Truscott that he was worried about the lack of combat experience among American soldiers and that he wanted Truscott to arrange for as many Americans as possible to have this opportunity with the British Combined Operations Headquarters. Truscott left this meeting deeply impressed with Marshall, a man he greatly admired for the rest of his life. He was also very clear about his new assignment, later recalling 'I was never under any illusion ever as to what General Marshall wanted me to do.' At this stage, however, Marshall was not that impressed with Truscott. Shortly after Truscott's interview, Marshall went to Eisenhower and asked him if he was sure that Truscott was the right man for this job. It required skill, diplomacy and intelligence, and Truscott had come across as a being too rough and raw. Eisenhower stood his ground and Marshall backed his subordinate's judgement. In time Marshall would change his mind about Truscott's abilities.

Working at Lord Louis Mountbatten's headquarters

Before leaving for his new assignment, Sarah and the youngest child Jamie moved to Charlottesville, Virginia, to stay at the family home there. Upon arrival in the United Kingdom Truscott reported to Lord Louis Mountbatten who 'welcomed us gracefully'. While recognising Mountbatten's ease of manner and personal charm, Truscott sensed in him other, less desirable character traits. 'He is quite a person,' Truscott wrote to Sarah. 'I see in him a hardness and perhaps a selfishness or overweening ambition that I have heard no one mention.' Mountbatten's Combined Operations Command had been organized from personnel from the three services and had control of the commando units as well as the responsibility for planning the cross-channel invasion. Truscott soon learned that Mountbatten's organization was not popular with the other services:

> During the summer that I was in Mountbatten's headquarters there was quite a bit of jockeying around. It was quite evident that Mountbatten and his headquarters was not nearly as popular among his own services as it was with the PM. But there was quite a deal of pulling and hauling.

This was a natural effect of the various services having to lose some of their best men in order to form the commando units. Added to this was the fact that the commandos had the opportunity for direct action while the regular forces continued training for it. The commandos also received a great deal of publicity which further increased the resentment felt by the regular formations. Two weeks after his arrival in the United Kingdom Truscott was promoted to brigadier general. He immediately noticed that his elevated rank made things smoother at Mountbatten's headquarters, especially an increased willingness to co-operate and share information.

Despite the problems experienced, Truscott and his staff decided that the best way for American soldiers to gain combat experience was to form their own commando unit and attach it to Mountbatten's command. Truscott wrote a proposal to form this new unit and forwarded it to General Marshall on 26 May 1942. The proposal was approved two days later. There was an issue as to what to call this new elite unit which would be a battalion of around 500 men. 'Commandos' was an inherently British name, and the US senior commanders wanted a uniquely American one. Truscott looked back to US frontier history and a unit of irregular fighters attached to the British Army during the Seven Years War called Rogers' Rangers. The new organization would be called Rangers and the men for it were initially taken from the 34th Division and the 1st Armored Division, which at that time were the only American divisions then in the United Kingdom, though neither was yet at full strength. The men applying to join the Rangers had to be fully trained, skilled in the use of their weapons, motivated to fight and physically fit, which included having 20/20 vision. The man selected to lead this elite unit was an impressive 33-year-old West Point graduate named William Orlando Darby. Freshly promoted to the rank of major, Darby had just ten days to create this new unit.

Planning Operation Torch

Concurrent with planning the Dieppe raid, Truscott was given a new assignment. He was to leave Mountbatten's command but remain in London to work on the Operation Torch landings being planned for French North Africa. General Marshall had selected Patton to command one of three invasion forces in the operation. On 9 August Major General George Patton arrived in the United Kingdom to learn all he could about Operation Torch. The senior officer who briefed him on the latest iteration of the plan was none other than his old polo-playing friend stationed in London and working at a British headquarters. Patton was pleased to receive such a thorough briefing and at the end of it said: 'Dammit, Lucian, you don't want to stay on any staff job in London with a war going on. Why don't you come with me? I will give you a command.' Truscott replied that he would like nothing better but that the final decision where he served was up to Eisenhower. Patton then formally asked for Truscott. Eisenhower, who knew the qualities of both men, readily approved Truscott's transfer to Patton's command.

First amphibious landing: Dieppe

Before returning to the United States to take up his new role, Truscott experienced his first amphibious landing, albeit as an observer. This was Operation Jubilee, planned and executed by Mountbatten's Combined Operations Command, the landing at Dieppe on 19 August 1942. Truscott informed Patton prior to the latter leaving London that he had been offered an observer role in Operation Jubilee. Patton not only approved of Truscott going along as an observer but wished that he could have been included. Later Mark Clark refused permission for Truscott to observe the large raid, but Eisenhower overruled him. It tainted Truscott's relationship with Clark, which would take some time to improve. Four officers and 44 men of Darby's 1st Ranger Battalion were involved in Operation Jubilee as direct participants. They were split into two groups and were to attack the beaches on the flanks of the main landing with the British commandos.

For a hit-and-run raid, Operation Jubilee was a large affair involving 250 ships, 73 landing craft and just over 6000 men. It was also a disaster with over half the men who landed at Dieppe being killed or taken prisoner of war. Among those killed in action were five Rangers, including two officers. Truscott, who had found the Dieppe raid 'a novel and thrilling experience' never considered the raid a failure as so many others did. He later wrote: 'It was an essential though costly lesson in modern warfare.' Truscott had also made history by being the first American general of the Second World War to see combat in Europe.

Although he spent the day aboard a British destroyer, the *Fernie*, watching events unfold, Truscott did admit to Sarah that: 'I have seen war — and have been in danger — and have seen men die on the land, in the sea, and in the sky.' What Truscott did not tell Sarah was that a large shell had hit *Fernie* killing or wounding sixteen sailors and that a large metal nut from the ship had struck Truscott's boot.

Sitting in a large wardroom surrounded by wounded men, Truscott felt in desperate need of a cigarette. He took out his tobacco pouch and rolled a cigarette, lit it and took a drag. A voice nearby asked if he had another cigarette to spare. Truscott recalled: 'I gave him my cigarette and then, until my sack was empty, I rolled and lit cigarettes for the others.' It is hard to imagine other generals with this common touch. The Dieppe raid, costly failure that it was, did achieve one decisive outcome. It presented incontrovertible proof how difficult amphibious operations were and that the Allies were nowhere near ready to invade France.

First operational command: Operation Torch

Truscott returned to the United States in September to work as Patton's deputy. The plan for Operation Torch had been finalised with Patton to command the Western Task Force of some 35,000 men. It would sail directly from the United States and land on Atlantic Ocean beaches in North Africa. Its objective was Casablanca in French Morocco, held by forces of Vichy France, and the key airfield of Port Lyautey. A Central Task Force commanded by Major General Lloyd Fredendall would land at Oran in Algeria while the Eastern Task Force under US Major General Charles Ryder had the city of Algiers as its objective. Patton split his Western Task Force into three sections, each with an assigned mission. He aimed to capture Safi 130 miles south of Casablanca, Fedala just north of the city and the airfield at Port Lyautey some 80 miles northeast of Casablanca. Truscott was placed in command of the assault on Port Lyautey, codenamed Operation Goalpost, and to achieve it he had the 60th Infantry Regiment Combat Team, an Armored Battalion Combat Team and 40 smaller units of infantry, artillery, tank destroyers, engineers, medical personnel and sundry other units. This was a force of just over 9000 officers and men and more than 1000 vehicles including tanks and towed artillery. The heart of it though was the infantry regiment and armoured battalion. Port Lyautey was actually a river port some nine miles up the shallow and meandering Sebou River. The capture of the airfield was the prime mission. Trying to organize a realistic rehearsal for Operation Goalpost caused Truscott considerable problems. The navy was unco-operative, restricting the rehearsal to one small beach that had no obstacles. When Truscott appealed to Patton via a phone call to intervene it earned him a stern rebuke:

> 'Dammit Lucian, I've already had enough trouble getting the Navy to undertake this operation. All I want is to get them to sea and to take us to Africa. Don't you do a damn thing that will upset them in any way.'

With that, Patton slammed the phone down. While Patton knew and liked Truscott, this incident raised some doubts in Patton's mind whether Truscott was up to the task ahead and if he could command men in battle. Patton wrote in his diary on the eve of the Moroccan landings: 'I am just a little worried about ability of Truscott. It may be nerves.' Patton did not yet know his man.

Truscott went around Patton and appealed directly to a senior admiral he knew and respected, one Vice Admiral Kent Hewitt. Through Hewitt he was able to have a full rehearsal on several landing beaches at night. They revealed some serious shortcomings which were soon rectified. It made Truscott reflect on his role and the importance of thorough preparation. He wrote of it: 'My own mistakes and the mistakes of others in preparing this command for battle would be paid for in the lives of Americans for whom I was responsible. It was a sobering thought.' It also established in Truscott's mind just how important thorough preparation, including a realistic rehearsal, was to achieving success in battle. He wrote: 'I had learned that preparation is the first essential for success in war, and that the adequacy of preparation reflects the capacity of a commander and his staff.'

On 23 October Patton's Western Task Force sailed from the United States heading for French North Africa. The journey of some 4500 miles would take just over two weeks. Operation Torch, the invasion of French North Africa in November 1942, was, according to Rob Citino, the Allies' 'first great amphibious invasion of the war'. The Dieppe raid, which Truscott had witnessed that year, showed 'just how difficult such an operation could be'. Operation Torch was to be no different.

Arriving on the evening of 7 November, Truscott's landing at Port Lyautey did not start well. To take the airfield, which was the primary objective, first the town of Port Lyautey and its adjacent fortress known as the Kasbah had to be captured. Truscott's force could then push onto the airfield which was in one of the river's loops on high ground north

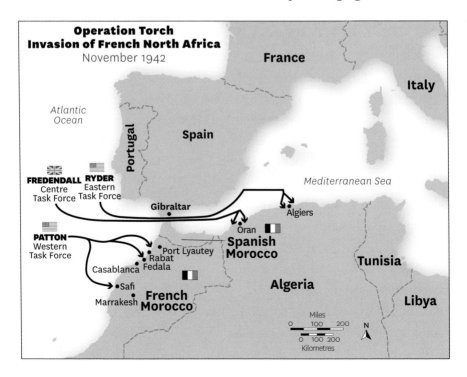

of the town. Truscott had planned to land his force on five separate beaches, aiming to surround the town and Kasbah before sailing upriver in an old destroyer and capturing the airfield. Truscott's plan did not even survive contact with the North African shoreline.

The first problem was that the naval commander could not pinpoint where they were on the African coast and valuable time was lost trying to ascertain if they were in the right location. To add to this, not all of the ships carrying the 60th Infantry Regiment had turned up. Truscott decided to launch the attack with the one battalion that had arrived, but once it was ashore at dawn on 8 November, he lost all contact with it. Pacing the deck of the ship trying to remain outwardly calm, Truscott was both irritated and frustrated at the lack of information he was receiving. He decided he needed to find out for himself.

Truscott went ashore while it was still dark to see what was happening only to find a very confused situation. No officers were present at the command post on the beach, just some sleeping soldiers and radio operator whose radio was dead. Truscott sent out two aides to find out what was happening and sat down on a sand dune to wait. Sitting on that dune, he had never felt more alone in his life and must have wondered if this was the end of his army career.

It very nearly was. On shore Truscott encountered a tank destroyer and two tanks with their crews confused about where they should go. He ordered them to accompany him. As they advanced inland an accidental burst of one of the tanks .50-calibre machine guns passed just inches above his head, nearly decapitating him. Later Truscott returned to the command post on the beach still no wiser about how the landing was progressing. He later recalled: 'more than anything else right then I wanted a cigarette'. Violating his own blackout order, Truscott lit up. Instantly hundreds of soldiers lit up around him 'as other lonely and uncertain men sought the comfort of tobacco'.

Despite all the detailed planning for Operation Torch many things had gone wrong, and Truscott's landing was no exception. Arriving at the west coast of North Africa, the navy transports lost formation so not all the troops arrived in time for H Hour. Heavy seas made disembarkation difficult, causing further delays. Some troops were landed miles from where they should have been and many of the

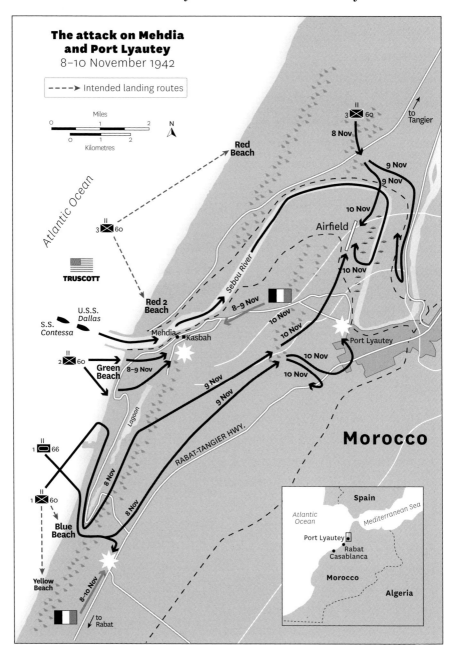

French defenders put up a fierce resistance. If all had gone according to plan the airfield at Port Lyautey and the port and nearby towns of Sale and Mehdia would have been captured by sunset on 8 November. None of this happened and it took Truscott's force two days of hard fighting before the airfield and towns were in their hands on the morning of 11 November. The fight to capture the Kasbah was particularly vicious and Truscott had to call in close air support from eight navy dive bombers before the fort was captured.

Hours later the airfield was also captured by a scratch force from three battalions, several tanks and two 105 mm self-propelled guns. Shortly after the airfield's capture, a general ceasefire went into effect across French North Africa. The next day, 12 November, Truscott wrote to Patton blaming the navy for many of the problems he experienced at Port Lyautey:

> We had a hell of a war up here, primarily due to the fact that the navy was one hour and forty-five minutes late in landing us and had some confusion as to beaches. Further, when we failed to take the battery by assault, owing to the loss of surprise, the navy took station about halfway to Bermuda and consequently the unloading progressed very slowly.

Allied losses in Operation Torch were relatively light when compared to future amphibious landings. Just under 1200 Allied soldiers were dead or missing and more than 1000 had been wounded. Truscott's losses in Operation Goalpost were 80 men killed with a further 250 wounded. At Port Lyautey the French buried around 300 of their soldiers killed in action. The Allies had been lucky. These actions in North Africa showed how inexperienced they were in amphibious landings and that they had much to learn. If the defenders had been better armed and determined to resist the invaders, Operation Torch could have been a disastrous and costly failure.

Eisenhower's deputy

Eisenhower and Patton were pleased with Truscott's performance during Operation Torch. In his officer efficiency report Patton described Truscott as a 'superior organizer and trainer, and a superb fighting leader of men'. Patton rated Truscott as Number 4 out of the 183 generals he knew. Little wonder that Truscott was promoted to major general on 21 December. Patton was somewhat uncomfortable that Truscott now held the same rank as him, writing in his diary: 'His promotion has been well deserved and he has invariably done a good, though never brilliant, job. I am very proud of him.' Proud or not, Truscott's elevation to major general meant that he could no longer remain in the Western Task Force. Truscott then reported to Eisenhower in Algiers eager for a new assignment.

It was not until 29 December that Truscott received one. He was to establish a small command group near the First Army headquarters at Constantine in north-east Algeria and act as Eisenhower's deputy. It was a staff position but one close to the front line where he could witness the action. Truscott was not in this role long, although he was Eisenhower's eyes and ears during the Allied defeat at Kasserine and Sbiba in Tunisia in mid-February 1943. He recalled of this first clash of American and German arms:

> That was the bitter picture which I had to convey to Eisenhower. More than one hundred tanks destroyed in two days, along with two battalions of artillery overrun, and two battalions of infantry lost, and no one knew how much more.

Truscott recommended to Eisenhower that the American general in command of the US II Corps, Lloyd Fredendall, be relieved of command and replaced by George Patton. According to Truscott, Major General Lloyd Fredendall 'had lost the confidence of subordinates', had not commanded II Corps well in Tunisia and did get on well with the British whom he thought could not be trusted. It was a damning if honest assessment and Eisenhower implemented Truscott's recommendation to replace Fredendall with Patton.

Less than three weeks later Eisenhower assigned Truscott to a new command of his own. It was another plum assignment. Major General Lucian Truscott Junior was to replace Major General John Lucas as the commander of the 3rd Infantry Division. It would not be the only time that Truscott was called in as Lucas's replacement. To echo Patton's assessment above, Truscott had performed well but not brilliantly during Operation Torch. This new appointment as the commanding officer of an infantry division was the chance Truscott needed to prove just how good he was.

3 Commanding the dogface soldiers

The new commander

Command of the 3rd Infantry Division was a notable appointment for Truscott, one that would make or break his military career. The 3rd Infantry Division was a regular army formation that had been activated in November 1917 in North Carolina while Truscott was undergoing his cavalry training. The division had distinguished itself in July 1918 when Paris was threatened by a German offensive, earning itself the label as the 'Rock of the Marne'. After the First World War, the division was one of the few infantry formations that the US Army did not deactivate. Its shoulder patch, a square with blue and white diagonal stripes, was later made famous by war hero Audie Murphy in his book and film *To Hell and Back*.

Truscott assumed command of the 3rd Division on 8 March 1943. He spent the first week getting to know his divisional staff and officers and watching the men train. Then, as one of the battalions was to move to a new position some ten miles north of Port Lyautey, he issued an order that it was to march at four miles per hour instead of the regulation two-and-a-half miles an hour. Truscott later explained his rationale:

> I had formed strong views concerning the standards which should be expected of American infantry divisions in war. . . . I had long felt that our standards for marching and fighting in the infantry were too low, not up to those of the Roman legions nor countless

examples from our own frontier history, nor even to those of Stonewall Jackson's 'Foot Cavalry' of Civil War fame.

This new rate of marching was the standard set for commandos and Rangers and Truscott believed it should apply to American infantry too. At first there was some resistance from the battalion commanders who stressed that this rate was too high for American infantry. In the first attempt at this new rate of marching more than a hundred men from a total of about 1000 fell out and could not keep up. However, the second battalion that tried had just experienced some tough training in North Africa's sand dunes for two weeks. In the march back to base at Truscott's new rate only twelve men fell out. Truscott persevered and the standard he wanted was surpassed within two weeks of solid training. Eventually the standard for 3rd Infantry Division marching, combat loaded, was five miles in the first hour, four miles an hour for twenty miles and three-and-a-half miles for distances up to 30 miles. This strenuous marching pace became known throughout the division as 'the Truscott Trot'.

A later initiative of Truscott's was to adopt 'The Dogface Soldier' as the division's official song. During the Second World War the term 'dogface' became a nickname for the American soldier. It was a term of endearment when used by soldiers but an insult if used by anyone else, especially by other services like the navy or Marine Corps. In 1942, two members of the US Army Air Force wrote 'The Dogface Soldier' which became instantly popular with US soldiers. When Truscott heard it, he liked it and decided that this would be the 3rd Division's song. Singing together had always been part of the cavalry tradition and Truscott was immensely fond of it. Its opening lines went:

I wouldn't give a bean
To be a fancy pants marine
I'd rather be a dog faced soldier like I am.

The song concluded with:

> So, feed me ammunition
> Keep me in the 3rd Division
> Your dog faced soldier boy's okay.

It proved to be an excellent choice. The soldiers of the 3rd Division loved their official song. They sang it, marched to it and even danced to it when it was played at social functions. 'The Dogface Soldier' made it on to the pop chart in 1955 when it was used in the film *To Hell and Back*, a biopic about and starring Audie Murphy who had served with the 3rd Infantry Division. The song, with some of the lyrics altered, remains the official march of the US 3rd Infantry Division. According to H. Paul Jeffers, it is 'the only song known to have been promoted by a U.S. major general'.

There is little doubt that Truscott soon stamped his authority on the 3rd Infantry Division and that the standards he set transformed it into the best division in the US Army. This was certainly the opinion of Eisenhower who, after an inspection of the army formations under his command, wrote to General Marshall that:

> From every indication it is the best unit we have brought over here. Truscott is the quiet, forceful, enthusiastic type that subordinates instinctively follow. If his command does not give a splendid account of itself, then all signs by which I know how to judge an organization are completely false.

Truscott was proud of what he achieved with the 3rd Division prior to its deployment to Sicily. He recalled in 1959 that it was the greatest satisfaction of his military career, telling an interviewer that:

> 'The 3rd Division, when it went into Sicily, was probably the finest fighting aggregation that our Army has ever had. They were in

magnificent condition, and in such a condition that when they got into a fight they didn't even know they had been there. There wasn't another unit in the Army that could move like they could move. It was phenomenal.'

Preparing for Operation Husky

Receiving notice that the 3rd Division was to be one of the lead formations in Operation Husky, the invasion of Sicily, Truscott initiated a new training cycle. Husky was Truscott's third amphibious landing, and he was determined to avoid the confusion and chaos that had dominated the previous two. Part of this training involved his division spending time at the new Invasion Training Center at Arzew, Algeria, run by Brigadier General John W. O'Daniel whose well-earned nickname was 'Iron Mike'. Here the 3rd Division soldiers improved their fitness through speed marching, running obstacle courses, rope climbing and calisthenics. More importantly, they participated in several beach landings, made by both day and night, and against defended positions. They also learned other necessary skills such as fighting through a town's streets and how to advance behind an artillery creeping barrage. Truscott was so impressed with the training programme at the Invasion Training Center that he later appointed 'Iron Mike' O'Daniel as his assistant divisional commander. O'Daniel joined 3rd Division just prior to the Sicily landings.

The soldiers of the 3rd Division also attended other specialist schools that dealt with various topics such as how to handle landing craft, how to maintain communications, beach organization, equipment maintenance and mountain warfare. Knowing how important a good plan was, Truscott established a planning board within 3rd Division. As well as having representatives from the usual staff sections within the division, Truscott added others that would be invaluable during an amphibious landing. These included representatives from the navy, artillery, armour, Rangers, engineers

Major General Lucian Truscott Junior just after his appointment as commanding officer of the 3rd Infantry Division. The photograph was taken in Bizerte, Tunisia, in early July 1943, just prior to embarking for the invasion of Sicily. Truscott transformed the 3rd Infantry Division into one of the best infantry formations in the United States Army. The shoulder patch Truscott wears is that of the 3rd Infantry Division. The three white stripes on a blue background represent the number of the division and the three major operations the 3rd Infantry Division took part in during the First World War. One of these earned the division the title of the 'Rock of the Marne', which it still uses today.

Some senior Allied officers inspect an invasion task force off the coast of North Africa. The task force was for the invasion of Sicily and the officers seem concerned about what they are seeing. The three officers in the foreground are, from left, General Sir Harold Alexander, Lieutenant General George S. Patton and Rear Admiral Alan G. Kirk.

and even a commando-trained British officer. The planning board met daily in a designated 'war room'. Truscott wrote of this initiative: 'It was effective because it insured the utmost in co-operation among all the branches and services involved and the careful and co-ordinated planning of an infinite number of details.' Clearly Truscott had learned much from the chaos of his Port Lyautey landing.

One vital piece of information missing from the planning board was aerial photographs of the intended landing beaches. Senior planners at Eisenhower's headquarters ignored Truscott's frequent request for them until finally refusing point blank to have any taken for fear of giving away Allied intentions. Truscott then approached the US Army Air Force Eighth Bomber Command to see if they had taken any images of Sicily's beaches. Their commander, Major General James H. 'Jimmy' Doolittle, informed Truscott that while his command had no photographs on hand, he would have some taken. A few days after making his request Truscott not only had the images he needed but Doolittle also provided an aerial-photograph interpreter to assist his planning staff. Thanks to Truscott's initiative and Doolittle's 'can-do' attitude, Truscott's planners had, in his words, 'far superior photographic intelligence for this operation than any other Sub Task force — far better than I was ever to have again'.

The training for the 3rd Division culminated in a full-scale dress rehearsal that was so realistic that many of the soldiers thought the real invasion was underway. This rehearsal was critical, Truscott believed, to instilling confidence in his soldiers that the navy could do the job and land at the right place at the right time. Historian Russell Weigley has written that when 3rd Division landed at Licata in Sicily it was 'an operation characterized by divisional staff planning of meticulousness exceptional even by the standards of an amphibious assault'. Truscott's previous experience with amphibious landings had made a critical difference.

Commanding Joss Force on Sicily

Operation Husky was a complex amphibious operation in size larger than Operation Overlord, the cross-channel invasion of France. Launched on 9 July 1943, Husky involved the British Eighth Army landing on the eastern and southern shores of the island while Patton's newly formed Seventh Army landed on the southern coast in the island's centre. Truscott's 3rd Division, part of II Corps commanded by Lieutenant General Omar Bradley, was on the Seventh Army's left flank. Truscott had under his command a Ranger Battalion and a combat command of 2nd Armoured Division as well as his own 3rd Division. It was designated Joss Force, and its task was to capture the port of Licata and a nearby airfield by the evening of the first day. It was then to extend the beachhead to link up with II Corps on its right. Above

COMMANDING THE DOGFACE SOLDIERS | 55

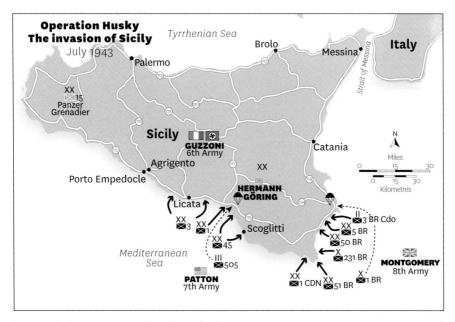

US Infantry soldiers advance through the narrow streets of Scoglitti in Sicily and begin their advance across the island. Operation Husky, the amphibious landing on Sicily's southern coast, was very successful. Truscott's 3rd Division soldiers found the actual landing to be easier than training for it.

all it was to protect Seventh Army's left flank from counterattack at vulnerable points. In a military first, Truscott's Joss Force would travel the 90 miles from the shores of North Africa in landing craft. This was an advantage in that troops did not have to clamber over cargo nets at sea into their allocated landing craft. It also complicated matters as three types of landing craft were used, and they each travelled at different speeds.

For the first time in Truscott's experience, despite a slight delay in the landing time, everything went according to plan. The planning and rigorous training was clearly evident. The landings, which had not been preceded by a naval bombardment, achieved complete surprise. The four landing beaches were quickly cleared of all resistance and within an hour Truscott had ten battalions, a Ranger battalion and supporting tanks ashore and moving inland at speed. All of its objectives were captured within seven hours including Licata and the nearby airfield. Over 1000 prisoners of war were taken, with Truscott's casualties numbering little more than a hundred. The detailed planning and intense training had clearly paid off. Several officers and soldiers told Truscott that 'fighting the battle was a damn sight easier than training for it'.

Other landings had not gone so well, but Patton had 50,000 men and 5000 vehicles on Sicily's shores by the end of the first day. By nightfall of 10 July, contact had been established along a 50-mile front from Licata to Scoglitti, with more men and supplies due to land in the coming days. The Seventh Army started to advance inland with the ultimate goal of capturing Palermo, the largest and most important port in Sicily. Truscott, always towards the front of his division and often under fire, was instrumental in the capture of this town.

Palermo, on Sicily's northern coastline, was meant to be captured by Montgomery's Eighth Army with Seventh Army protecting its left flank. Such a passive mission was not to Patton's liking, and he coveted the glory that would come with the capture of such an important military objective. As a prelude to taking Palermo, Patton, bypassing Bradley,

Lieutenant General George S. Patton in March 1943. Patton was Truscott's commanding officer in North Africa and his army commander in Sicily. While the two men were old polo-playing friends, this did not prevent Patton from threatening to relieve Truscott of command for what he perceived as a lack of aggression and undue caution. Despite this, Truscott enjoyed serving under Patton and was uncomfortable with replacing Patton as commander of the Third Army at the end of the war.

authorised Truscott to capture Porto Empedocle on the southern coast further west of the Seventh Army boundary but to pass it off as a reconnaissance in force towards Agrigento. On 14 July Truscott used a Ranger battalion, the 7th Infantry Regiment and seven battalions of field artillery with 148 guns to encircle Agrigento and then attack the port. Both Agrigento and Porto Empedocle were captured that day along with 6000 Italian prisoners and hundreds of tanks and guns. The risk had paid off and four days later the theatre commander, General Harold Alexander, after a visit from Patton extolling the virtues of the mission, unleashed Patton's army towards Palermo. Patton was not going to play second fiddle to Montgomery any longer.

Onto Palermo and Messina

Capturing Palermo was not going to be easy. It was over a hundred miles away to the north-west through some rugged mountain terrain. The 'three tortuous roads northward had steep grades, numerous hairpin turns, and many bridges which would be rendered more difficult by enemy demolitions and delaying actions,' Truscott later recalled. Truscott briefed his officers on 18 July that Palermo was now the new objective and that he expected them to be in the city in five days. He then produced a bottle of scotch whisky and proposed a toast: 'To the American doughboy.'

Truscott's 'doughboys' or 'dogface soldiers' as he preferred to call them set off for Palermo, sometimes covering more than 30 miles a day through blistering heat and choking dust. A private in 15th Infantry Regiment recorded in his diary: 'We are walking at the rate of 4.5 miles an hour. Boy are my dogs [feet] barking now.' Truscott's 3rd Division and the Ranger Battalion captured Palermo on 22 July. Instead of taking five days as he had predicted, Truscott's dogface soldiers had reached the city in just three. It was a stunning victory, one that captured 53,000 Italian prisoners of war and left the Allies in control of half the

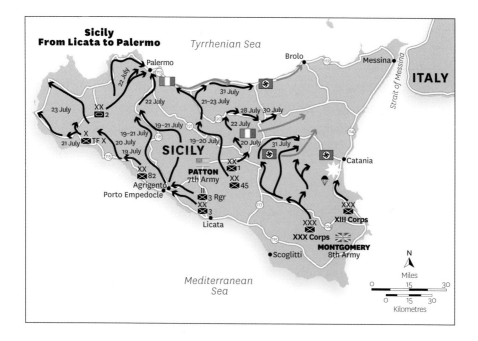

island. Patton was elated. When he met up with Truscott in Palermo, he slapped him on the back and said, 'Well, the Truscott Trot sure got us here in a damn hurry.'

The Palermo victory was not enough for Patton, and he wanted more of the glory that came with success. He now set his sights on racing Montgomery to Messina on Sicily's north-eastern tip. Over drinks in the Royal Palace on 26 July, Patton confessed to Truscott that he would 'certainly like to beat Montgomery into Messina'. Three days earlier, on 23 July, recognizing that Montgomery's Eighth Army would need help taking Messina, General Alexander authorised Patton to advance from the west towards the city. Patton's Seventh Army advanced eastwards along two parallel roads: Highway 113 along the north coast and Highway 120, the inland road. For this drive to the east the 1st and 45th Divisions would be the lead with Truscott's 3rd Division in reserve. As his army moved east, Patton became obsessed with reaching Messina before Montgomery's Eighth Army. He informed Bradley, commanding II Corps, that if he had to lose men in order to reach Messina a single day ahead of Montgomery then the men must be sacrificed. Bradley was shocked. On 28 July Patton spoke to Troy Middleton, commander of the 45th Division, stating: 'This is a horse race, in which the prestige of the U.S. Army is at stake. We must take Messina before the British.' In fact, there was no race. Montgomery was focusing on winning the battle he was currently fighting, a tougher struggle than he anticipated. He had agreed with Alexander that it would be best for Patton's Seventh Army to take Messina.

After Middleton's 45th Division captured a ridge at Santo Stefano along the coast road, Truscott's 3rd Division took over the advance. Truscott's division advanced around ten miles from the San Stefano ridge before striking serious opposition at another significant ridge. This was the step ridgeline at San Fratello where the German 29th Panzer Grenadier Division was well established. A series of probing attacks failed to dislodge the Germans and Patton could sense he was losing the race. On 10 August he ordered Bradley to mount an amphibious 'end run' by landing a battalion on the north coast twelve miles behind enemy lines to seize Monte Cipolla near Brolo. Capturing Monte Cipolla would sever Highway 113, entrapping the 29th Panzer Grenadier Division and giving Truscott's division a clear run to Messina, now just 40 miles away. Patton wanted the amphibious landing made the next morning.

When Bradley ordered Truscott to undertake this new amphibious landing, Truscott asked for an extra day to prepare it. This was to enable him to arrange adequate artillery and to have additional infantry ready to support the landing. Bradley, who considered this mission 'trivial' and possibly 'foolhardy', readily agreed to this delay. Patton was furious when he learned that Truscott had requested more time. The situation was not helped when Patton's deputy, Geoffrey Keyes, mistakenly informed him that Truscott did not want to undertake this new amphibious operation. Patton was outraged and informed Truscott by phone that the operation would happen. Still seething with anger, that evening Patton rushed to Truscott's

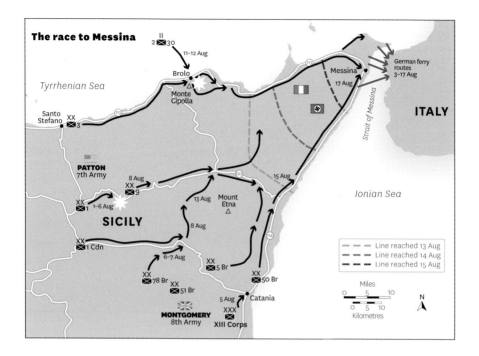

command post, and, in the words of Rick Atkinson, there occurred 'another of the ugly confrontations that bedeviled Patton across Sicily'. It almost cost Truscott his command.

Patton found Truscott pacing with a map in his hand and immediately launched into him. Their conversation, during which Truscott refused to back down, is recorded below:

Patton: Goddammit Lucian. What's the matter with you? Are you afraid to fight?

Truscott: General you know that's ridiculous and insulting. You have ordered the operation and it is now loading.

Patton: General Truscott, if your conscience will not let you conduct this operation, I will relieve you and put someone in command who will.

Truscott: General, it is your privilege to reduce me whenever you want to.

Patton:	I don't want to. You are too old an athlete to believe it is possible to postpone a match.
Truscott:	You are an old enough athlete to know that sometimes they are postponed.
Patton:	This one won't be. Remember Frederick the Great: *L'audace, toujours l'audace* [Audacity, always audacity]. I know you will win.
Truscott:	General, the operation is loading. But I will tell you one thing, you will not find anyone who can carry out orders which they do not approve as well as I can.

Having made his point, Patton's anger evaporated. He put his arm around Truscott's shoulder and said: 'Damn it, Lucian, I know that. Come on, let's have a drink of your liquor.' After the drink Patton returned to his headquarters and reflected in his diary: 'I may have been bull-headed.'

The new amphibious landing did not go well. A depleted infantry battalion, just 650 men, lacking artillery, air and infantry support, was cut to pieces on the day of the landing. It lost 167 men for negligible effect. The German panzer grenadiers withdrew on 12 August and Patton was elated. Reaching Mount Cipolla, which was littered with dead American soldiers and mules, he stood up in his staff car, polished metal helmet gleaming in the sun. He pointed to the hill with his swagger stick. 'The American soldier is the greatest in the world,' he proclaimed. 'Only American soldiers can climb mountains like those.'

Lucian Truscott always believed that the extra day he wanted for this mission would have made a considerable difference. He did not hold this against Patton, reflecting after the war that Patton 'was a very fair man. I liked serving under him.' Truscott remained 'a great admirer' of Patton for the rest of his life. The corps commander, Omar Bradley, whose star was on the rise and who would later be Patton's commanding officer, reacted differently. He later remarked:

'Patton arrogantly overruled me . . . For the sake of a favorable headline, Patton was placing the lives of many men in jeopardy.' Bradley never forgave Patton for his meddling in his command and for ordering the unnecessary sacrifice of men just to beat Montgomery into Messina.

It was on the road to Messina that the distinguished war correspondent and Pulitzer Prize-winning journalist Ernie Pyle first came across Major General Lucian Truscott. Pyle later wrote a vivid and sympathetic account of the man he found at the front line of the action. A German demolition was preventing his division from moving on Messina and Truscott was checking the progress being made to repair the road. Pyle wrote of his introduction to Truscott:

> During the night Major General Lucian Truscott, commanding the Third Division, came up to see how the work was coming along. Bridging that hole was his main interest in life that night. He couldn't help any, of course, but somehow he couldn't bear to leave. He stood around and talked to officers, and the men. After a while he went off a few feet to one side and sat down on the ground and lit a cigarette. A moment later a passing soldiers saw the glow and leaned over and said, 'Hey give me a light, will you?'
>
> The general did, and the soldier never knew he'd been ordering the general around.

Pyle then described how Truscott stretched out among some rocks and took a short nap. He was rudely awakened when an engineer, dragging an air hose, got it tangled around Truscott's feet. The engineer snapped at Truscott in the dark: 'If you're not working, get the hell out of the way.' Truscott, 'without saying a word' moved further back. Pyle's account reveals much about Truscott. He was up at the front, taking the pulse of the action. If men were doing their jobs and trying to get things moving, he stayed out of their way and let them get on with it. If they were inactive or lacked purpose and direction, he was there to provide it, often accompanied by the sharp edge of his tongue. This was a general who led from the front.

On 16 August 1943, four days after the landing to take Monte Cipolla, the 3rd Division overcame the last German resistance and reached the heights overlooking Messina. The next day when the senior Italian officer surrendered his pistol to Truscott, Truscott told him to keep it and present it to General Patton later. Patton arrived at 10 a.m. on the morning of 17 August and just after Patton's entourage entered Messina a British armoured patrol arrived in the city from the west. Patton had won his race by just a few minutes. Truscott and his staff celebrated into the night and the alcohol flowed. An aide recorded that: 'A very nice time was had by all. Scotch highballs, cognac and champagne enjoyed after the mad dash to Messina.'

Patton's triumph would be short-lived. On 21 August he was forced to apologize to two soldiers he had slapped while they were convalescing in hospital. This was followed shortly afterwards by Patton having to apologize to every unit in Seventh Army. Patton's Seventh Army was then deactivated, and he would not receive another active command for almost a year. It would not be the last time that Patton would be relieved of command of an army.

Before he was relieved of command Patton had to write officer efficiency reports on all his senior officers. Regarding Truscott, Patton wrote: 'I know of no major general who has more efficiently performed as a Division Commander.' He ranked Truscott Number 5 out of 153 general officers. There is no doubt that Truscott had performed well on Sicily and his 3rd Division had been the outstanding formation in Patton's Seventh Army. As mentioned above, Truscott had enjoyed serving under Patton, a commander he deeply admired. He could not say the same about his next army commander.

To the Italian mainland

On 3 September 1943 the Allies invaded the mainland of Italy with Montgomery's Eighth Army landing on Italy's toe at Reggio. Six days later, on 9 September Montgomery made another landing at Taranto on Italy's heel the same day that Mark Clark's Fifth Army landed at Salerno on Italy's west coast, south of Naples. The newly created Fifth Army was the first American Army Headquarters to be created outside of the United States and the Salerno landing would be its baptism of fire. While Eighth Army had an easy passage, Clark's landing was fiercely resisted and almost driven back into the sea. At one point, the situation was so desperate that Clark considered evacuating the beachhead.

During the Salerno landing, Truscott's 3rd Division, which was considerably understrength due to the recent actions on Sicily, was held in reserve at Palermo in Sicily to rest and refit. At a briefing on 5 September, Clark had indicated to his senior commanders that he expected little resistance at Salerno. He told Truscott to ready his division for another amphibious landing somewhere in the north. The experienced Truscott did not think Salerno was going to be easy and he followed Clark's progress after the Salerno landing with increasing alarm. Sensing that it would soon be needed, Truscott sent a message to Clark that he was coming to Italy and requested a meeting. Truscott met Clark on the morning of 15 September and advised him that the 3rd Division at Palermo could load for Italy as soon as ships were available. Clark replied that he needed the 3rd Division as soon as possible and that it would be assigned to VI Corps. Truscott left at once to arrange the move. To make up the 2000 soldiers his understrength division needed, replacements were taken from the 1st and 9th Divisions. On 20 September the entire division was ashore at Salerno. By this time the beachhead was secured, providing the springboard to advance north to Naples. But already Clark's casualties were more than 13,000 with some 2000 being killed in action and a further 7000 wounded. It was a terrible start to the campaign.

The Allies were now fighting in a terrain that favoured the defender. Shaped like a boot, Italy runs north to south for 750 miles, but its waistline is narrow, ranging from 80 to 120 miles. The Apennine Mountains run almost the full length of Italy, incorporating many mountains, peaks, hills, ridges and ravines while hundreds of rivers and streams are required to drain this high ground. There are only two narrow coastal plains, one on each side of the country. The climate is harsh with baking hot, dry summers and winters of rain, wind, mud and snow. This was a terrain where wide outflanking manoeuvres were impossible, where air power was inhibited by weather and where the kingpin of the battlefield was not heavily armoured tanks but well-trained infantry skilled in mountain warfare. From Naples, the Fifth Army's first significant objective, just two roads ran north to Rome. This was Highway 7, the Appian Way, which ran along the west coast, and Highway 6, the Via Casilina, an inland route. The Germans controlled both roads and were well prepared to defend them.

Crossing the Volturno

The 3rd Infantry Division's first task was to be part of VI Corps' advance to the Volturno River twenty miles north of Naples. The advance was difficult because of the poor, narrow roads and torrential rain that started to fall on 26 September. Enemy resistance was primarily through the use of demolitions defended by a rearguard. They were easily brushed aside but did slow the advance. Truscott's tactics were to have one of his infantry regiments advance along the road until it struck a serious obstacle. This was usually a bridge that had been destroyed or a huge crater blown in the road. The Germans were adept at creating these obstacles and they always covered them with fire. While one battalion maintained contact with the enemy, pinning them in position, the other two battalions would attempt to outflank the Germans. Truscott ensured that the advance battalion always had a forward artillery observer attached who could call in artillery

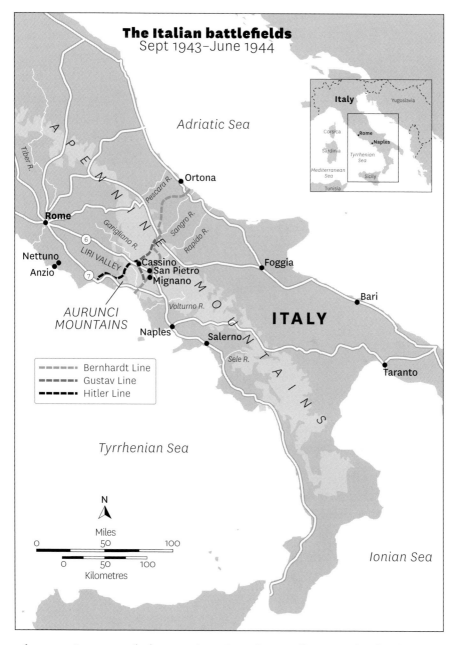

whenever it was needed. Sometimes it took just a few rounds of well-directed artillery fire to dislodge the defenders. Then Truscott's engineers could come forward and repair the damage or remove the obstacle.

Progress was steady but slow. Naples was taken on 1 October 1943 and the Allied soldiers pushed on to the flooded Volturno River where the Germans had prepared their next defensive position. Here the Fifth Army would attempt its first crossing of a contested river. As Rick Atkinson has noted, the Volturno was 'a formidable obstacle'. To make the crossing, Fifth Army moved six divisions to the riverbank on a 40-mile front where they faced four divisions of the German Tenth Army.

In the centre of the Allied line, Truscott had concealed his division for almost a week prior to the attack. His division's sector was seven miles long and the river was swift and wide at some 200 feet. It was also six feet deep in places with steep ten-foot banks. Being in flood, the river flowed fast enough to knock heavily laden solders off their feet. On the night of the attack, listening to the rumble of artillery with flashes in the night sky, the VI Corps commander Major General John Lucas wrote a poignant entry in his diary:

> Tonight we crossed the Volturno. Have been working on it for days. I have done all I can and now I'm in the hands of God and my subordinates. A solemn thought [is] that your name on an order means the death of many men.

Truscott was one of the subordinates Lucas was depending on. The 3rd Division's guns, mortars and heavy machine guns opened up at 1 a.m. on the morning of 13 October and almost an hour later Truscott's infantry, carrying life vests, rafts, small boats and anything else that would float, slid down a muddy ten-foot embankment and into the cold waters of the Volturno. Heavy smoke saturated the riverbanks, offering some protection from observation. Truscott's 3rd Division soldiers easily crossed the river and by dawn five battalions were across it. They did not linger but pushed on to Monte Caruso four miles beyond the river. But Truscott's infantry were vulnerable. The British on their left flank had not been able to cross the Volturno, leaving 3rd Division's left flank exposed. Also, no armour or heavy support weapons were across the river.

That morning Truscott walked along the riverbank urging speed and action. He directed his artillery to keep firing where they thought the Germans were likely to be and then personally led his engineer platoon down to the raging river. His direction to them in his gruff voice was inspirational: 'You've only got picks and shovels, men, only your hands, but right now they're better than guns. For God's sake, let's get the job done. We've got a whole regiment of men over there. They'll get wiped out unless you get tanks across.'

The engineers did not need to be told twice. They ran to the river and started breaking down its steep banks with shovels, picks and bare hands. It took them some hours, but they did it. US Sherman tanks began fording the river at 11 a.m. Truscott later described his engineer's actions that morning as 'immediate and inspiring.' His own performance could also be called that.

Still Truscott urged more action. A tank commander puzzled at the rapping on the hull opened the turret to find a two-star general yelling at him. 'Goddamit, get up ahead and fire at targets of opportunity. Fire at anything shooting our men, but goddamit do some good for yourselves.' An engineer officer who complained of the difficulties of bridging the river also earned a Truscott tongue lashing. 'What do you mean it can't be done? Have you tried it? Go out and do it!' It was done and, in the afternoon, the first light bridge was across the river. Massed, accurate artillery fire protected the infantry across the river and soon the entire Volturno valley was in Allied hands and the bridgehead across the river was expanded the next day.

The crossing of the Volturno was a major Allied success. Moving quickly on a broad front and crossing this major obstacle advanced the Allied line 35 miles beyond Naples. While many formations and units had been involved in crossing the Volturno, none had done more to achieve this success than Truscott's 3rd Division. Major General John Lucas, relieved that his first major action as a corps commander had been successful, wrote in his diary after a visit from General Harold Alexander: 'The big Chief [Alexander] seemed immensely pleased with what we are doing and expressed his great admiration for the American soldiers.'

Success at the Volturno had come with a high price, with some 314 men being killed or wounded. And looking up from their new positions the Allied infantry soldiers could see a range of hills looming ahead. Behind these were more hills and rivers and Rome was still 130 miles away. This was no friendly terrain or climate for the dogface soldiers.

Italy's mountains and mud

Truscott's 3rd Division remained in the mountains and mud of Italy for 59 days. Conditions were appalling. Heavy rain, biting cold, thick mud, no hot food, limited supplies and high rates of sickness, including malaria, jaundice and trench foot, mirrored conditions of the First World War. The situation was aggravated by the fact that Fifth Army soldiers had not been issued winter clothing, an appalling omission. The soldiers suffered in a climate that chilled them to the bone.

The 3rd Division's last action prior to being withdrawn in preparation for the Anzio landing was an attempt to take Mount Lungo. For ten days the division had been struggling in the Fifth Army centre to capture Monte la Difensa, one of the mountains in a narrow valley known as the Mignano Gap. Progress was slow, the terrain difficult and conditions were horrendous, but Truscott was confident they would soon capture the feature. Then, on 5 November, Truscott received an unexpected phone call from the corps commander Lucas. Truscott was ordered to withdraw his division immediately, load it on motorized transport that evening, move to Rocco Pipirozzi, about twenty miles away, and attack north-west in the early morning to capture Mount Lungo. Lungo was an isolated hill mass in the British area of responsibility. Truscott was shocked at receiving such a peremptory order and protested to Lucas that he needed to perform a reconnaissance first and arrange adequate artillery support. Lucas, sounding both stressed and anxious, was

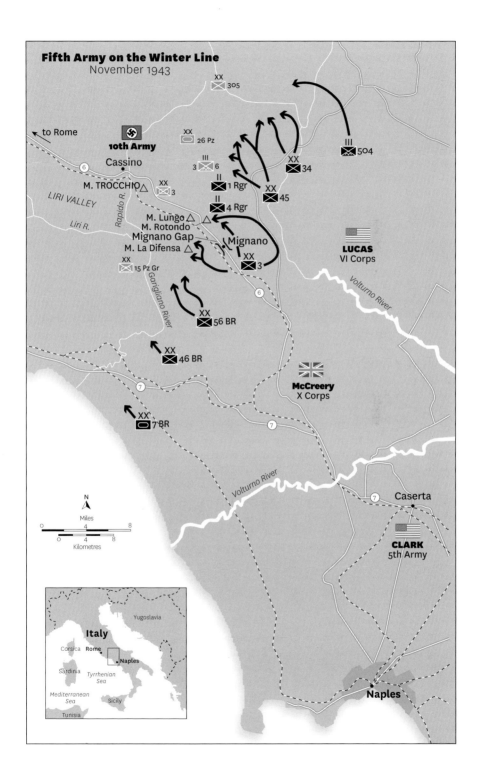

An impromptu conference in the early morning of 6 November 1943 in the Cassino sector, Italy. Truscott, at left in the bathrobe, confers with his army commander, Lieutenant General Mark Clark, who is holding the other half of the map. The officer between them is Lieutenant General Courtney H. Hodges who was then commanding the (inactive) Third Army. Despite the smiles, Truscott and Clark had a difficult relationship, especially after Truscott was promoted to command VI Corps.

adamant that it had to be done right away. Lucas explained that Clark had boasted to the British that Truscott's 3rd Division could take the feature. Clark's pride was at stake. When Truscott asked to speak to General Clark directly, Lucas replied with some desperation: 'No Lucian, damn it. You know the position I am in with him. That would only make it worse and put me in a helluva hole. You have just got to do it.'

There was nothing more to be said and Truscott agreed to move the 3rd Division that night. However, he told Lucas he thought the chances of success were slim. Lucas agreed but directed Truscott to do the best he could.

Truscott's best was good, but it was not enough. At 0530 hours on 6 November his 30th Infantry Regiment attacked and fought their way over across the spurs of the north slope of Mount Cesima. They came within sight of the next mountain, called Rotondo, but could get no further. The next day the entire 3rd Division went into the attack and after bitter fighting, much of it hand to hand, the division captured Mount Rotondo and gained a foothold on the southern end of Mount Lungo. From there the soldiers could see another mountain, Monte Cassino, looming ahead of them. But the 3rd Division, like all the divisions then fighting in the Mignano Gap, was now a spent force and could get no further. It spent the next ten days holding this newly won position against repeated counterattacks. On 17 November 3rd Division was relieved by the 36th Division and moved into reserve where they would rest and replenish for the next mission. Little wonder that Truscott later recalled: 'This last operation which began on October 31, was to be a heartbreak for me.'

Since arriving in Italy two months earlier the 3rd Division had been constantly in action, the price it paid for being so good. Its casualties had been heavy and had reached 8600. The losses included almost 400 officers, including over half of the division's young second lieutenants. Nearly 4000 privates had become casualties and the three infantry regiments, which had done the bulk of the fighting, had all lost 70 per cent of their strength. Truscott felt these losses deeply and wrote to a fellow officer about 'how fragile an infantry division really is'. He also wrote to Sarah: 'I only pray that I can live up to what my lads seem to expect of me.'

On 18 November, the first day out of the front line, Truscott issued a message that he ordered be read to every soldier in his 3rd Division. It summarized their recent performance in Italy and praised all involved:

> This record is one in which every officer and man can feel just pride. It has been made possible only because of your sound training and discipline, your unexcelled physical condition, the splendid leadership of the officers and non-commissioned officers, and most important the magnificent loyalty and determination which has driven you to every objective regardless of hazard and difficulty.

It was praise that had been well-earned and Truscott deserved it too. However, the 3rd Infantry Division's next assignment, another amphibious landing this time almost 90 miles behind the enemy lines at Anzio, would be full of hazards and difficulty. It would test the qualities of the dogface soldiers and the skill of their commander to the limit.

4 Disaster at Anzio

Planning Operation Shingle

By early 1944 the Allies were stalled at Cassino and were being forced to make direct frontal assaults on the Gustav Line where the German defences were strongest. They needed to try something different. By this time, as Rob Citino has written, 'the Anglo-American alliance had a great deal of experience at amphibious landings, and it was almost inevitable that they decided to try one here in Italy.' This was Operation Shingle, an attempt to outflank the Gustav Line by landing at Anzio in the German rear just over 30 miles south of Rome. D-Day for Operation Shingle was to be 22 January 1944.

The concept of an amphibious landing well behind the Gustav Line at Anzio was both bold and risky, given that the Germans were bound to respond vigorously to any threat to their rear. The logistical requirements were considerable, not least because sustaining a bridgehead enclave between the Gustav Line and Rome would entail a great deal of effort. The landing also proved controversial as the Allies and their senior commanders had different and opposing views as to what Operation Shingle was to achieve. The British leader Winston Churchill and the theatre commander Alexander envisaged a shattering blow that would break the deadlock in Italy by outflanking the Gustav Line and possibly even capturing Rome.

The army commander Mark Clark initially endorsed this concept, but his experience at Salerno made him more cautious. Instead, he settled for a solid beachhead, one that reached to the Alban Hills which would provide a solid foundation for further exploitation. His final advice to Major General John Lucas, the officer commanding the VI Corps which would undertake the landing, was: 'Don't stick your neck out, Johnny. I did at Salerno and got into trouble. You can forget this goddam Rome business.' Lucas was pessimistic about the chances of success at Anzio, writing that: 'This whole affair has a strong odour of Gallipoli.' At a conference of senior commanders to discuss Shingle, Lucas had felt 'like a lamb being led to the slaughter'. He did not need Clark's advice urging caution. For Lucas it was enough to get VI Corps ashore intact, secure a beachhead and hold it against the counterattacks that were sure to come. Only when the beachhead was secure would VI Corps move further inland.

In reality, Lucas's small force did not permit much more than the establishment of a shallow beachhead. The limited number of landing craft retained in Italy after Operation Husky was just 68. This was enough for only two divisions, with some supporting Rangers and commandos, for the initial landing. More men and equipment could be put ashore if the initial landing succeeded, but such a small force could not be expected to take the Alban Hills or march on Rome. Operation Husky on Sicily had used two complete armies, while Shingle involved only one corps, with the initial burden falling on just two divisions. These were the British 1st Division (commanded by Major General Ronald Penney) landing to the left of Anzio-Nettuno and the 3rd US Division (under Truscott) on the right. Five battalions of Rangers and commandos and four battalions of paratroopers would be part of the initial landing, but it was still a small force. Landing a few days after D-Day would be part of the 1st Armored Division and one regiment from the US 45th Division. The US official naval historian, Samuel Morrison, wrote of the Anzio landing: 'That was the fundamental weakness of Operation Shingle. Either it was a job for a full Army, or was no job at all; to attempt it with only two divisions was to send a boy on a man's errand.'

Truscott concurred with this assessment, later writing:

> No one below army level believed that the landing of two divisions at Anzio would cause a German withdrawal on the southern front [Cassino] or that there was more than a remote chance that the remainder of the 5th Army will be able to cross the Rapido River and fight its way up the Liri and Sacco valleys to join us within a month.

The stage was set for some of the hardest conditions ever experienced by Allied soldiers. In the words of Harvey Ferguson: 'Thus began a four-month operation that would resemble the closest thing to the Great War's trench warfare occurring during World War II.' Truscott would need all his command skills and a considerable degree of luck to survive the ordeal ahead.

The final decision to undertake Operation Shingle was made on 8 January 1944 in Marrakech, Morocco, at a meeting of senior military leaders and Winston Churchill. This left just two weeks for final plans and preparations. Lucian Truscott had been uncomfortable with many aspects of the Anzio landing, especially the lack of time to prepare for it. Mark Clark's initial plan had been for just one division, Truscott's 3rd Division, to land with enough supplies to last seven days. Truscott was appalled and expressed this forcefully to Clark:

> We are perfectly willing to undertake the operation if we are ordered to do so and we will maintain ourselves to the last round of ammunition. But if we undertake it, you are going to destroy the best damn division in the United States Army for there will be no survivors.

At the end of 1943, Truscott's 3rd Division had moved to a training area near Naples. In the first week of January the division began preparations for another amphibious landing. Knowing how difficult these operations were, and that success depended on the army and navy working in complete harmony, Truscott insisted on carrying out a full-scale rehearsal. In this he was supported by VI Corps commander Major General Lucas. The rehearsal took place on the Salerno beaches on the night of 17 January and, although it was not a complete fiasco, much went wrong. The assault battalions had disembarked too far at sea, some fifteen miles offshore, so that few landed on the correct beaches. Those that did landed late. No artillery, tanks or tank destroyers had been landed so that the advancing infantry were on their own. A considerable amount of vital equipment had been lost at sea including trucks, howitzers, radio equipment and several men. Chaos dominated the landing beaches, and the original plan was in disarray. Only the assault battalions reached their objectives, primarily due to the experience and leadership of their officers and non-commissioned officers. However, against an opposition skilled in defence and counterattack these battalions would have been wiped out, with the landing being nothing but a costly disaster. The rehearsal for the British division had gone better but even that had failed to disembark either a brigade or divisional headquarters.

Truscott was furious about how little time was available to prepare for Operation Shingle. He sent a terse message to Clark's Chief of Staff, Major General Al Gruenther, asking: 'If this is to be a forlorn hope or a suicide sashay, then all I want to know is that fact.' With Lucas's blessing, Truscott also wrote a detailed account of the rehearsal, stressing all that had gone wrong and demanding another one. Clark replied:

> Lucian, I've got your report here and it's bad. But you won't get another rehearsal. The date has been set at the very highest level. There is no possibility of delaying it for even a day. You have got to do it.

Having made his point, Truscott prepared his division as best he could in the limited time left to him.

As the troops of the 3rd Division boarded their transports in the Bay of Naples, the division's band played a series of military marches. The soldiers in formation marched to their waiting ships with guidons flying and their officers saluting Lucian Truscott watching from the deck of the USS *Biscayne*. The dogface soldiers were silent until the band struck up the division's anthem played in march time. Then they all burst into song:

> I wouldn't give a bean
> To be a fancy pants marine
> I'd rather be a dog faced soldier like I am.

Watching from the ship, Lucian Truscott was both pleased and proud. He stated years later that: 'It may not have been the best tradition from a security point of view, but it was one of the most inspiring things that ever happened to me.' It indicated too that morale was good in his 3rd Division. Despite the limited time for preparation, Truscott knew his division could be relied upon and could overcome any difficulties ahead of it.

Operation Shingle: the landing at Anzio

The place designated as the landing zone for Operation Shingle was around the seaside resort towns of Nettuno and Anzio barely 30 miles south of Rome. Anzio in ancient Roman times had been called Antium and was notorious as the birthplace of emperors Caligula and Nero. Legend had it that it was in Antium's theatre that Nero fiddled while Rome burned. Nettuno, the smaller of the port towns, was just over a mile north-east of Anzio.

In the early hours of 22 January 1944, Lucian Truscott's fourth amphibious landing was underway. It commenced at 1.50 a.m. with a short, intense barrage by two naval rocket craft along the intended landing beaches. The water was calm and the weather forecast good. Anchoring the left flank of VI Corps, the British 1st Division and two battalions of commandos from the 2nd Special Service Brigade landed five miles north of Anzio. On the right flank, Truscott's 3rd Division landed just south of Nettuno, while Darby's Rangers and a parachute infantry battalion headed for the casino building overlooking Anzio Harbour. The landing was virtually unopposed and caught the German defenders by surprise. Soon more than 200 tired and dispirited Germans were prisoners of war.

The Allies had been helped by the misleading assessments of the Abwehr, the German military intelligence agency. The Abwehr had assured Field Marshal Albert Kesselring, the German commander in Italy, that there was no reason to suspect the Allies could mount another amphibious operation in his theatre of war. Its intelligence chief, Admiral Wilhelm F. Canaris, had informed Berlin headquarters that: 'There is not the slightest sign that a new landing will be undertaken in the near future.' Senior commanders in Italy were then advised by Kesselring on 15 January that 'a large-scale landing operation [was] out of the question for the next four to six weeks'.

General Lucas recounted watching the landing from a ship off the Anzio coast: '... could not believe my eyes when I stood on the bridge and saw no machine gun or other fire on the beach'. Everything had fallen into place for this landing. Enemy resistance was slight, the navy had found the correct beaches, the sea was calm and the weather fine. It meant most soldiers landed only wet to the knees and eager to advance inland. Truscott was pleasantly surprised that the landing had gone so well. He wrote of it: 'After the almost disastrous performance during the rehearsal, our navy comrades gave us one which was unbelievably smooth and accurate.'

The region around Anzio was relatively flat, most of it being reclaimed marshland. From Anzio, a good road and railway ran due north along a ridgeline through the town of Aprilia, soon called 'The Factory' by Allied soldiers because its visible bell tower resembled an industrial smokestack. From Aprilia the road ran to the town of Campoleone, an important road and railway junction, and from there to Albano which was eighteen miles from the coast. Albano was on the southern slope of the Alban Hills, also known as the Colli Laziali, the dominant feature in the region. At Albano, this northern road joined Highway 7 (the Appian Way), the main coastal road from the south to Rome. Around Albano the country was broken by numerous deep ravines. This high ground was drained by several rivers including the Moletta and Astura, which reached the sea in the Allies' new beachhead. The area directly north of Nettuno was flat farmland that had been reclaimed from the Pontine Marshes. It was drained by a series of canals, but even in dry weather the water table was usually just two feet from the surface. It did not take much rainfall for this area to flood. Two main roads passed through Nettuno. One road followed the coast east to the town of Littoria, beyond which it joined Highway 7. The other road ran north-east from Nettuno through the towns of Cisterna and Cori to reach Valmontone on Highway 6, which was the main road from Naples to Rome. In the months ahead many of these towns would become deadly battlefields.

But not on the first day. By its end Lucas's invasion force of 36,000 men with some 3200 vehicles were ashore. Darby's 1st, 3rd and 4th Ranger Battalions and the attached 509th Parachute Infantry Battalion had easily captured both Anzio and Nettuno by 7.30 a.m. Further progress that day established a significant beachhead of some fifteen miles in width and from two to four miles in depth. Just thirteen Allied soldiers had been killed in action. Forming a semicircle from the sea, the left flank was on the Moletta River with the right extending to the Mussolini Canal. Lucas resisted the temptation to push on to the Alban Hills. Instead, he focused on holding the port of Anzio, installing anti-aircraft defences and constructing an airfield. Both Alexander and Clark visited the beachhead just after 9 a.m. on that first day and they were pleased with what Lucas had achieved. They also endorsed Lucas's decision not to push further inland. As one wit recorded of the visit of Clark and Alexander: 'They came, they saw, they concurred.'

While everything had gone well on Day 1, it was still a difficult day for Lucian Truscott. He had gone ashore at 6.15 a.m. mute with laryngitis and with an inflamed throat. Seeing that things were progressing well and still feeling miserable, Truscott took the opportunity to have a short nap in a thicket near the beach. After inspecting his division's front lines and finding nothing amiss, he returned to his command post where he hosted Generals Clark and Lucas and several staff officers to breakfast. Clark complimented Truscott on the performance of his 3rd Division and the Ranger Battalions. As Rick Atkinson has written, 'Poor Truscott croaked his thanks.'

The Germans respond

While the landing on 22 January had caught the Germans by surprise, they responded with their usual alacrity. Within a few hours of the landing Kesselring had a full report and quickly realized that this was not just a raid. Kesselring admitted that the first hours of 22 January 'were full of anxiety' but he soon recognized the weakness in the Allies' tactics. He was pleased to learn of the 'hesitant advance of these troops', the lack of armour and the small size of the landing force. As he later wrote: 'It was a half-way as an offensive . . . a basic error.' Kesselring recognized too that the Alban Hills were the key to any quick advance on Rome and that he needed to secure them before the Allies did. By 8.30 a.m. German reinforcements were being rushed to the Anzio-Nettuno area. They came from Rome, Northern Italy, France, the Balkans and Germany. Unfortunately, only a trickle of these reinforcements came from the Cassino front, which had been one of the pressing reasons for the Allies to make the landing in the first place. Within six hours Kesselring had all or parts of eleven German divisions converging on the Alban Hills. Hitler urged Kesselring to drive the enemy into the sea and directed that: 'The battle must be waged with holy hatred.'

On the night of 22 January, the Hermann Göring Panzer Division arrived and immediately attacked the Allied right flank. It drove the leading elements of 3rd Division back over a canal and captured some Americans. The next day the Hermann Göring Panzer Division was itself driven back. It withdrew to the town of Cisterna, which it immediately fortified.

Building up the force

Over the next few days, Lucas focused on building up his beachhead force and its logistics base of food, ammunition, equipment and supplies. Only limited advances were made. On the left flank, the British 1st Division edged forward and captured the town of Aprilia after driving out a battalion of panzer grenadiers of whom a hundred were taken prisoner. By 25 January the division was halfway to Albano and the Highway 7 junction. On the right, Truscott's 3rd

Division also edged forward. It struck heavy German resistance three miles south of Cisterna and could not make further progress. Truscott and his staff worked out plans for a full divisional assault on Cisterna and also one for the British to take Campoleone, but Lucas refused to sanction it. He did not feel ready to launch such large-scale assaults until he had more forces available. He especially wanted a combat command from the 1st Armoured Division to be available, which was expected to arrive in the next few days. The attempt to take both towns would be delayed until 30 January, during which time another German division arrived at Cisterna to reinforce the Hermann Göring Division defending the town.

In the afternoon of 24 January Truscott was wounded during a German air raid on the beachhead. An anti-aircraft shell exploded six inches from Truscott's left foot and several fragments hit him in the leg. Had Truscott not been wearing his long and thick cavalry boots the wounds would have been serious. As it was, an army surgeon had to remove several fragments from his leg before wrapping the leg in an adhesive cast. Against his wishes, Lucas awarded Truscott a purple heart. Along with Truscott's raging sore throat and laryngitis, the wound was just one more impediment to endure. He soldiered on, now limping as well as speaking in a hoarse whisper.

By 25 January, the third day of the landing, Lucas's force on the beachhead numbered 56,000 soldiers and nearly 7000 vehicles. The force continued to grow and soon doubled in size to include the 1st Armored Division, the 45th Infantry Division, a parachute infantry regiment and the 1st Special Service Force of Canadian and American commandos. Both Clark and Alexander visited the beachhead on 25 January and were still pleased with what had been achieved. Clark advised Lucas to prepare for a large German counterattack but directed that he should secure both Campoleone and Cisterna as soon as possible in order to provide a firm base for further exploitation. Lucas agreed to do so, but that night he confided in his diary: 'I must keep my feet on the ground and my forces in hand and do nothing foolish.

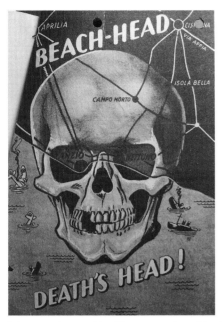 

'The Beach-Head had become a Death's Head!' warned a German propaganda leaflet dropped to Allied soldiers at Anzio. While the Allies suffered over 40,000 casualties in the Anzio beachhead, Germany's losses there were similar.

This is the most important thing I have ever tried to do, and I will not be stampeded.' The need to do nothing foolish and his refusal to be stampeded would soon cost Lucas his command.

A disastrous advance

Lucas did not launch the assaults on Campoleone and Cisterna until 30 January, a full week after the initial landing. By then Kesselring had 70,000 German soldiers ready to resist any expansion of the beachhead. Alexander and Clark were now disappointed with Lucas's caution while, in London, Winston Churchill fumed at the inaction. The plan was to launch an infantry assault after an intensive air, artillery and naval bombardment. When it finished the British 1st Division on the left would take Campoleone. When it held the town, the US 1st Armored Division was to advance beyond this town and capture Albano, cut Highway 7 and seize the Alban Hills. On the right Truscott's 3rd Division, Darby's Rangers and the 504th Parachute Regiment would attack Cisterna, push up Highway 7 to Velletri and press on to Valmontone and cut Highway 6. As Harvey Ferguson has written: 'It was an ambitious plan.'

Too ambitious by far. The British attack on Campoleone was a costly failure and the 1st Armored Division lost 24 tanks just reaching the British position. Their commander, Ernest Harmon, went forward the day after the assault to see if he could assist the British infantry. He recalled a horrible sight: 'There were dead bodies everywhere. I had never seen so many dead men in one place. They lay so close together that I had to step with care.' The British attack had carved out a salient some four miles deep and two miles wide but had failed to capture its prime objective of the junction town of Campoleone.

On the right, Truscott's attack had fared worse. Allied intelligence had not detected the arrival of the additional German division at Cisterna so that instead of facing one weakened division along a broad front, Truscott's infantry and Rangers faced two. Eleven German battalions defended Cisterna, three times more than expected.

Truscott's plan, prepared with Colonel William Darby, had been for two Ranger battalions to infiltrate between strongpoints into the town of Cisterna. Once in position behind the German lines the Rangers were to create as much damage as possible. While this was happening, Truscott's 15th Infantry Regiment and a third Ranger battalion was to attack the German front and break through to the Ranger battalions in the town. While this main assault was happening, the 504th Parachute Regiment would make a diversionary attack on the Mussolini Canal, while Truscott's 7th Infantry Regiment attacked on the left along another canal aiming to cut Highway 7 north-west of Cisterna. This main assault would have support from artillery, tanks and tank destroyers.

The assault soon floundered. The two infiltrating Ranger battalions reached to within half a mile of Cisterna when they were detected by German paratroopers. They were soon being attacked by the full firepower of two German divisions and by the afternoon of the next day they had run out of ammunition. These two Ranger battalions had left with a force totalling 767 men. Only eight returned. Around 300 Rangers had been killed in action and more than 450 captured. These losses reduced their commander Colonel Darby to tears. The Germans later paraded these captured Rangers through the streets of Rome for a photo opportunity and to be abused by diehard fascists.

The attack by Truscott's infantry regiments had not gone to plan either. Little more than a mile of ground was captured on a five-mile front along with 200 prisoners of war. Truscott's infantry battled around Cisterna for two days but made little progress. The fighting since the landing had cost Truscott's 3rd Division 3000 casualties and a third of its tanks and tank destroyers. The Ranger battalion fighting with Truscott's infantry suffered 50 per cent casualties. After battling for two days and still over a mile away from Cisterna, Truscott's division was a spent force. Realizing this, Truscott gave the order to cease any further advance and dig in where they were.

When Lieutenant General Mark Clark learned of the extent of the Rangers' losses he was alarmed and was especially concerned about the bad press which would result. He wrote in his diary that using the lightly armed Rangers to spearhead 3rd Division's assault on Cisterna was 'a definite error in judgment'. Clark needed a scapegoat and settled on Lucian Truscott. Visiting the beachhead to meet with Lucas and Truscott, Clark accused Truscott of misusing the Rangers and that an investigation to determine responsibility for the disaster might be needed. Truscott was staggered and angry. He reminded Clark that he had been responsible for the creation of the Rangers, had worked with Colonel Darby on the plan and both of them knew full well their capabilities. He also added that there was no need for an investigation as the responsibility for the disaster was his alone. To his great credit Lucas intervened. He disagreed with Truscott as to where the responsibility lay. Lucas had seen Truscott's plan and approved it so ultimate responsibility was his. Knowing that he now had two scapegoats should the need arise, Clark let the matter drop.

German counterattacks

After the failure to take both Campoleone and Cisterna, the soldiers in the Allied beachhead prepared for the inevitable German counterattack. Intelligence from Ultra intercepts of German radio signals confirmed that Kesselring was planning a counterpunch. On 3 February Clark advised Lucas to abandon any further plans to advance. Instead: 'You should now consolidate your beachhead and make plans to meet an attack.' After softening the beachhead with several air attacks, the German counterattack was launched on 7 February. The local German commander, General Eberhard von Mackensen, had massed 96,000 troops, 100 tanks and more than 200 anti-tank and assault guns for the attack. It fell on the British positions on the left flank and aimed to eradicate the recently won salient around Campoleone. The fighting was bitter with heavy casualties on both sides. By the end of the day, the Germans had captured almost three miles of the salient and the British suffered 1500 casualties including 900 taken prisoner of war. The next day the rest of the salient was taken along with a further 800 prisoners. The Germans pushed on to take Aprilia, endangering the entire beachhead. At Aprilia, on 8 February the Germans easily beat off a counterattack by two US regiments from 45th Divisions supported by tanks. They consolidated their positions and prepared to launch an even larger attack, one ordered personally by Adolf Hitler and designated with the ominous name of *Fischfang* (Fishing).

In preparation for the German counterattack, Truscott had strengthened the front of his 3rd Division. He made extensive use of mines, barbed wire and obstacles. Despite the objections of their crews, Truscott moved his tanks and tank destroyers close to the front where they could immediately support the infantry. He grouped his seven battalions of artillery so that their fire could be co-ordinated by a forward observer. Instantly a threat was identified, all guns could fire on it. Likely targets were pre-registered, and communications tested. A standing order was issued that if communication was lost, the guns would keep firing on the last identified target until they were restored.

Operation Fischfang commenced in the early morning of 16 February. Adolf Hitler had ordered Kesselring to launch a 'concentrated, overwhelming, ruthless' assault on a narrow front using all the armour and artillery that was available in order to remove what he called the beachhead 'abscess'. Hitler ignored Kesselring and Mackensen's concern that such an attack over several miles of open country would leave the attackers vulnerable to concentrated artillery fire. They informed Hitler that the defenders had plenty of guns including fire from naval vessels off the coast. Kesselring and Mackensen planned to attack with six divisions and 200 tanks, including some of the latest Tiger and Panther models. Two more divisions, the 26th Panzer and 29th Panzer Grenadier Divisions, remained in reserve to exploit any success. Once across the three miles of open ground to the beachhead, the attackers could infiltrate through the scrub pines of the Padiglione Woods, split the beachhead in two and drive the defenders into the sea.

The morning of Wednesday, 16 February was cold and foggy, and a frost had hardened the ground. After a 75-minute artillery bombardment, the German infantry and tanks advanced. The attack had been expected. Aerial surveillance had detected the build-up and activity on the ground while Ultra intercepts confirmed where the attack was coming from and when. It fell predominantly on the British 56th Division on the left and the US 45th Division in the centre of the line. Despite being forewarned, the defenders suffered heavy casualties and lost just under a mile of ground. But the German attackers also lost greatly, some 1700 men on this first day, with the Infantry

Lehr Regiment, described by Hitler as 'elite killers', running away. Kesselring, alarmed at his dwindling artillery ammunition stocks, pressed Mackensen to commit his reserve, to be informed that: 'The time has not yet come.'

It was not far away, however. Under cover of darkness, German soldiers advanced along streambeds and goat tracks ready for another dawn assault. After a raid by Luftwaffe bombers, at 8 a.m. on Thursday, 17 February some 60 panzers and the infantry from three German divisions advanced down the Via Anziate, then pivoted to the east to strike the US 45th Division in the flank. By noon a wedge two miles wide and one mile deep had been driven into the centre of the 45th Division. One regiment lost 1000 men, most taken prisoner. The desperate situation was restored by Allied artillery including some naval gunfire and 800 planes providing the heaviest close air support of the war to that point. The beachhead survived, although it had shrunk by several square miles.

The climax of the battle came on 18 February, the third day of Operation Fischfang. Even though the beachhead was still intact, Mackensen felt that the time was right to commit the two divisions in reserve. As the panzer grenadiers advanced, these veterans of Sicily, Salerno and Cassino sang and shouted taunts. The demoralized defenders pulled back to the protection of the Allied guns.

Deputy Corps commander

By 18 February, with all eight divisions committed to the battle, the Germans had carved a bulge from the beachhead almost three miles deep and four miles wide. But the fire from the many artillery batteries defending the Allied perimeter held them at bay. On that day Lucas received a signal from Clark advising him that Truscott had been appointed as deputy commander of VI Corps. Truscott's beloved 3rd Infantry Division would now go to Brigadier General John W. 'Iron Mike' O'Daniel. Lucas knew that this appointment probably meant he would soon be replaced, recording in his diary that night: 'I think this means my relief in that [Truscott] gets the corps. I hope I am not to be relieved . . . I have done my best. I've carried out my orders and my conscience is clear.' It was a change of command that neither Lucas nor Truscott wanted.

Truscott's 3rd Division had easily beaten off the weak attacks made against it. When he saw the signal appointing him as Lucas's unwanted deputy, Truscott knew what it meant. Churchill, Alexander and Clark had lost confidence in Lucas and now his days in command were numbered. Pleased that in 'Iron Mike' he was leaving his 3rd Division in good hands, Truscott still hated leaving it. Writing to Brigadier O'Daniel in early April to congratulate him on the 3rd Division's role in thwarting the final German counterattack in February 1944, Truscott recorded his remorse at having to leave the Division: 'You know my regret at leaving the Division after our association of a year, even to command the Corps; however, I will always be, in spirit, a member of the Third Division.' This letter so moved O'Daniel that, against army regulations, he later presented Truscott with the 3rd Division's Combat Infantry Badge, stating that: 'No matter how poor your selection of branch at the very start may have been . . . that does not take your right to receive an award that this Division thinks you so justly deserve.' Despite his protests, Truscott was deeply touched by the gesture.

In spite of his reluctance to leave the 3rd Division, Truscott knew it was his duty to do so. He recorded of it: 'There was a job to be done, and I was a soldier. I could only carry out the order loyally.' But with a general like Mark Clark, loyalty was not enough. Clark craved success and reputation and he now turned to Truscott at Anzio to deliver it. If he failed there, Truscott's active military career, like that of Major General John Lucas, would be over.

5 Anzio command: Truscott's finest hour

Truscott as deputy commander

Truscott's appointment as Lucas's deputy on 18 February 1944 was a difficult time for both men. They had been friends for years and Truscott enjoyed serving under a commander he liked and respected. As Truscott later wrote, General Lucas 'had always sought and received my advice and recommendations, and had treated me with the utmost consideration'. At the same time, though, Truscott admitted that 'our methods of command were different. I was not blind to the fact that General Lucas lacked some of the qualities of positive leadership that engender confidence.' When Truscott reported to VI Corps Headquarters, he was surprised at the atmosphere of 'desperation' and 'hopelessness' that pervaded there. He recalled that 'the situation was far graver than we had realized in the 3rd Infantry Division'. Truscott tried to dispel the gloom by assuring the staff there that nothing was ever as bad on the ground as it looked on a map. It had little effect. Truscott then did something that Lucas seldom did. He went forward to the frontline areas to see how bad the situation really was.

At the front, Truscott met with Major General Ernest Harmon of the US 1st Armored Division and Major General William Eagles of the US 45th Division. Both were confident that the worst was over and that the German attack could be contained. Their views ran counter to the fact that the 45th Division had been mauled in the recent fighting and that the Germans had carved out a deep salient in the corps' beachhead line some four miles wide and deep. Truscott agreed with Eagles and Harmon, concluding that the time was right to launch a counterattack. That evening he suggested to Lucas that he mount an attack using all the corps' reserves and a newly arrived British brigade. Lucas was hesitant but was overruled by General Clark who was at the headquarters that evening and agreed with Truscott's assessment. Lucas reluctantly ordered the counterattack for the next day.

On the same day that he met with Eagles and Harmon, Truscott investigated why Operation Fischfang had almost succeeded. He concluded that the 45th Division had suffered more than it should have because it did not have enough artillery support. What little it had received dwindled to nothing once communications were cut and no request for fire support could be made. Truscott was determined this would not happen again. He met with Brigadier Carl Baehr, the corps artillery officer who had just arrived at the beachhead. At the meeting he introduced Baehr to the artillery commander of the 3rd Division, Major Walter 'Dutch' Kerwin. Truscott explained that he wanted the artillery fire of the entire corps organized and co-ordinated as they were in the 3rd Division. Kerwin's job was to visit all units and outline the system of fire control used in the 3rd Division. Baehr's role was to use his rank to ensure it happened. Truscott wanted the changes made that night. The following morning Baehr reported to Truscott that the changes had been made and that 'I have had the best lesson in artillery that I have had in thirty-five years' service in the artillery'.

It came just in time. Ernie Harmon, whose 1st Armored Division led Lucas's counterattack on 19 February, later recalled: 'The American artilleryman wrote a page in history at Anzio. His aim was deadly.' That morning Harmon's tanks set off after a barrage from 400 artillery and naval guns and supported by 200 US Army Air Force bombers. From foxholes lining the road, the infantry of 45th Division cheered as the tanks and infantry steamed forward, yelling 'Give 'em hell.' Harmon's tanks clawed back a mile of ground before running into intense fire from concealed German panzers defending a blown bridge. It was not until 1.30 p.m. that, with the bridge repaired, Harmon's tanks pushed on towards Aprilia. They made good progress until nightfall when Harmon recalled them. German casualties were heavy with some 200 panzer grenadiers taken prisoner. While the counterattack gained little ground, it succeeded in breaking the enemy's will to continue the offensive. Late that evening Kesselring proposed suspending Operation Fischfang and Hitler reluctantly consented. German casualties were heavy: some 5400 men. Casualties in VI Corps also exceeded 5000. With such heavy losses on both sides an uneasy stalemate settled over the beachhead.

There was one further high-profile casualty, the scapegoat for failing to achieve more at Anzio. The axe that had been hanging over Lucas's head fell on the evening of 22 February. At 8 p.m., in a private meeting, Lucas was informed that Clark 'could no longer resist the pressure' and he was relieving him of command. That pressure had come from many sources including Churchill, Alexander and the British commander at Anzio. Even Jacob Devers, the senior US military officer in the Mediterranean theatre, thought it was time for Lucas to go and for Truscott to take over. When Truscott was informed of the change, he protested that Lucas's sacking was unfair and likely to damage morale at Anzio. Clark abruptly dismissed Truscott's concern, stating that the decision had 'already been made'.

In many ways, Truscott was correct. Lucas's sacking was unfair. Major General John Porter Lucas was 54 years old in 1944. He had a reputation for integrity, fairness and a good tactical brain. General George Marshall rated Lucas highly as an officer of 'military stature, prestige and experience', and both Eisenhower and Patton had full confidence in him. Unfortunately for Lucas, he was highly sensitive to sending young men to die, cautious and he looked every inch his 54 years. As Robert Citino has written, 'even one of his staff officers noted that Lucas "never seemed to want to hurt anybody — at times, almost including the enemy," and Lucas himself once admitted that "I am far too tender-hearted ever to be a success at my chosen profession."' In the words of Rick Atkinson: 'Empathy might ennoble a man, but it could debilitate a general.' Despite this, all senior officers involved at Anzio, including Clark, Truscott, Templer, Harding and even George Marshall in the USA agreed that Lucas was correct to adopt a cautious approach at Anzio and focus on preserving the beachhead rather than striking for the Alban Hills and Rome. That evening, after learning of the change in command, Truscott went to Lucas to express his regrets:

> While Lucas was deeply hurt, he had no ill feeling toward me, and our friendship was unbroken up to the time of his death. But he was bitter toward General Clark and blamed his relief upon British influence. It was one of my saddest experiences of the war.

Truscott was later informed by Marshall that his appointment to command VI Corps at Anzio meant that he missed out on commanding a corps during Operation Overlord, the cross-channel invasion of France. Eisenhower had specifically requested him, but that would not be possible now that Truscott was needed at Anzio. Mark Clark needed Truscott to remain there. Clark explained in an interview shortly after the war that he chose Truscott because: 'He was a hard fighter. We had our little disagreements, but I had great confidence in his tactical ability and aggressiveness.' The senior American commander in the Mediterranean, Jacob Devers, told an interviewer in 1969 that 'the day that Truscott took over, things changed at Anzio'.

Truscott's changes

Truscott knew that the Anzio beachhead was far from secure and that he needed to make several changes in order to restore morale to the Allied soldiers fighting there. General Lucas had lost credibility in Alexander's eyes by remaining in his underground bunker and not getting out to the front to see things for himself. One of the first things Truscott did was to shift his main headquarters at Nettuno from the underground wine cellar that Lucas had used to a building above ground where he and his staff would also sleep. It was a symbolic move that showed the troops at Anzio that the corps commander and his staff shared their dangers. Truscott did not remain in his command post during the day but became a highly visible commander. During daylight hours he spent most of his time visiting the frontline units, seeing for himself what their status was and being seen by them. This was a complete change to what had previously occurred.

Truscott also changed his headquarters and the way they operated. He brought in several of his key staff from the 3rd Division to assist the existing staff with how the new general operated. One of these new ways was that Truscott wanted to be informed of any fresh development that occurred in the beachhead. He wanted this done by receiving a full and accurate briefing every morning at eight o'clock from section chiefs. This development kept Truscott and his staff fully informed about what was happening at Anzio.

Truscott had to do some coalition building with his British Allies who had been most vocal in pushing for Lucas to go. In this he was helped by receiving a senior British staff officer at his headquarters, Major General Vivian Eveleigh. As Truscott recalled: 'It was General Eveleigh who was primarily responsible for removing all previous causes of friction. His knowledge, boundless energy, and rare personal charm . . . were of inestimable value in establishing command relationships which was thenceforth to be a model for every Allied command.' It was also helped by Truscott's constant visits to British frontline units, which were appreciated by their commanders and soldiers alike. General Alexander visited the Anzio beachhead two days after Truscott took over command. When he returned to his own headquarters, Alexander told Lieutenant General Jacob Devers that: 'You have selected the right man to command that outfit. These [British] division commanders believe in Truscott absolutely and I don't think we're going to have too much trouble up there.'

Another change Truscott made was to improve the beachhead's anti-aircraft defences. Delegating the task to an able subordinate, Major General Aaron Bradshaw, with the assistance of the newly developed jam-resistant radar and using a standing barrage for aircraft that slipped under the radar, an effective air defence system was soon established. Within a month the anti-aircraft defence at Anzio had shot

Major General Lucian Truscott, the newly installed commander of VI Corps, confers with staff officers in the war room of his headquarters at Nettuno in early March 1944.

down or damaged over 400 aircraft. Losses for the Luftwaffe became so heavy that air strikes over the beachhead were abandoned.

German artillery, which outranged the guns used by the Allies, was soon nullified. General Baehr and his staff of the VI Corps Fire

Truscott, on the right, talks with legendary war correspondent and Pulitzer Prize-winning journalist Ernie Pyle, outside his war-damaged headquarters at Nettuno in the Anzio beachhead, in mid-March 1944. Pyle was later killed while covering the Pacific War.

Direction Centre developed an effective counterbattery plan that made artillery firing for the German gunners very dangerous. The moment a German gun opened fire it received accurate incoming fire from every corps gun within range. To help nullify the German monster railway guns dubbed 'Anzio Annie' and the 'Anzio Express', which the Allied gunners could not reach, two army chemical-warfare companies pumped out a giant smokescreen that protected the beachhead during daylight hours. All of these initiatives made a considerable difference to the Allied defenders.

Working with Mark Clark

Truscott often found Clark surprisingly tricky to work for and their relationship was never easy. Clark admitted: 'We had our little disagreements.' One such disagreement, so petty that it should never have been an issue, reveals much about the relationship. In his tours around the beachhead Truscott had noticed that all of the Fifth Army gasoline and ammunition dumps were concentrated in one area without any attempt at concealment. This was dangerous and the loss of these vital logistics could jeopardize the entire beachhead. He ordered the problem fixed and immediately earned a rebuke from Clark that he had no authority over these army personnel who were not part of his corps. Truscott replied that he thought all personnel in the beachhead were his responsibility and that perhaps these people should be part of VI Corps. If that didn't work, perhaps he should be designated as deputy army commander. Clark declined both suggestions. Instead, he sent a full colonel to be the Fifth Army representative on Truscott's headquarters with the authority to command army personnel in the beachhead. Truscott was fortunate that the colonel, Laurence Ladue, made himself part of the VI Corps team and had no problem implementing Truscott's directives, especially in relation to the fuel and ammunition dumps.

More German counterattacks

With another German attack expected, Truscott prepared his defences as best he could. One of the British Divisions, the 56th, had been severely mauled during Operation Fischfang and was down to a quarter of its normal strength. Truscott requested Clark to replace this division with one at full strength and send him reinforcements. Clark arranged for the British 5th Division to replace the 56th and promised to send the US 34th Division as soon as it was ready.

By 28 February, intelligence intercepts and aerial reconnaissance indicated that the Germans were about to launch another attack. Ordered by Hitler to make it, Kesselring and Mackensen decided to test the strength of the Allied right flank. Truscott believed that the

Germans would make a feint along the Via Anziante, while their main effort would fall on the 3rd Division's front south-west of Cisterna. He made every preparation possible to meet this threat. First, Truscott and his artillery chief Baehr mapped every conceivable German assembly point, likely routes that the enemy would use and the probable location of command posts. These were registered by the artillery for special attention. Truscott also arranged through Clark for an Allied bombing mission over the battle area. At 4.30 a.m. on 29 February, an hour before the designated start time of the German attack, every gun in the beachhead opened fire on their designated targets. The German attack proceeded exactly as the intelligence intercepts had indicated. Attacking with nine understrength divisions, five of which were directed against the 3rd Division's front south of Cisterna, they were met with a curtain of artillery fire, with the Allies firing some 66,000 rounds that day. The German attack petered out after suffering heavy casualties. Over half the tanks used in this attack were destroyed and the Germans lost 3500 men, having gained no ground at all. More than 1000 Germans were taken prisoner. A second attack on the next day had a similar result, after which Kesselring called it off. When Alexander visited 3rd Division in early March, he asked General O'Daniel how much ground his division had lost in this recent attack. He was impressed with Iron Mike's reply: 'Not a goddamned inch, sir.'

    Kesselring directed that all German units cease offensive operations except minor raids. The failure to eliminate the beachhead disturbed him. Kesselring's chief of staff, General Siegfried Westphal, had calculated that the Germans had expended more combat materiel in the attacks at Anzio than on any German battlefield since 1940, with the exception of Sevastopol, and with negligible results. This, Kesselring believed, was going to be 'the last year of the war' and Germany was going to lose it. For Kesselring, failure at Anzio was 'a turning point in the war' and he despatched Westphal to Berlin to inform Hitler and the German high command.

A counterattack planned by Truscott was cancelled when air support for it was unavailable. Both sides at Anzio needed to rest and regroup and now that bad weather prevented any further offensive operations. The situation at Anzio was deadlocked and would remain so for the next three months.

Truscott's health

When Truscott assumed command at Anzio on 22 February, he was not a well man, suffering from throat and teeth problems as well as limping from his recent wound. His condition was not helped by his excessive smoking and drinking. Regarding the latter, Rick Atkinson has written:

> Cognac, scotch, gin, and two ounces of rye before supper remain staples in the Truscott mess, but without evident impairment of the commander. He remained energetic, buoyant, and intensely absorbed in every detail of his beachhead domain.

In early March a specialist arrived at Anzio to give Truscott a full laryngoscopic examination. His condition did not improve and was further exacerbated by a hacking cough, not helped by the beachhead smokescreen, and nasal polyps. After another examination on 31 March, a doctor recommended voice exercises and a switch to cigars instead of cigarettes. Truscott stopped smoking for ten days but resumed the habit when he thought he was gaining weight.

Preparing the breakout

After making the preparations to break out from the beachhead, Truscott took the opportunity of the stalemate at Anzio to take five days' leave in Naples in late April. He returned in early May with a suntan and feeling much better. Noticing that the roses were blooming in abundance in the warm spring weather, Truscott had soldiers pick bouquets of them for the hospital wards of the 56th Evacuation Hospital. He wrote to Sarah from Nettuno that he felt younger than he should after 27 years of army service. He also added a note that revealed much about his personality, especially when compared to a general like Mark Clark: 'I hope that I have not become conceited or swell-headed, and I do not believe that I have. I have retained my sense of humour and I'm still able to laugh at myself.'

Determined to break the stalemate at Cassino and Anzio, the Chief of the Imperial General Staff, General Alan Brooke, sent one of his best officers to be Alexander's chief of staff. This was Major General Allan Francis Harding, known as John Harding. Harding was one of those rare officers who was an excellent combat commander as well as being a brilliant staff officer. It was no accident that he later reached the rank of Field Marshal. Harding studied the situation then existing in Italy and produced a plan that aimed to break open the Gustav Line and defeat the German forces in southern Italy. It was codenamed Operation Diadem.

Harding recognized that the three previous Cassino battles and the situation at Anzio had ignored a fundamental principle of war: the concentration of force at a decisive point. Instead of attacking with just one or two divisions, Harding's Operation Diadem would use the equivalent of three armies to break the deadlock. First, both the Eighth and Fifth Armies were concentrated on a twenty-mile front, leaving the Adriatic coast virtually denuded of Allied soldiers. Then the Anzio beachhead under Truscott's command would be reinforced to contain seven Allied divisions: the equivalent of a small field army rather than a corps. In Operation Diadem, the breakout from Anzio was to be part of a one-two punch. The Fifth and Eighth Armies were to advance in a parallel drive to the north-west, shattering the Gustav Line. Once they were beyond the Gustav Line and heading north, the breakout from Anzio would commence with the aim of trapping the retreating

German forces between the two armies advancing north-west and Truscott's enlarged VI Corps surging north-east from the beachhead. Part of Harding's plan was his use of deception. All military movements for Operation Diadem were made at night while false movements, military exercises and radio traffic convinced the Germans that a new amphibious landing of three divisions was being planned to the north of Rome. No such landing was planned, but the Germans started to prepare for it, keeping many of their reserve troops in the north to counter this new development.

Operation Diadem involved a massive concentration of force. The British Eighth Army contained six infantry divisions, three armoured divisions and three independent armoured brigades. Clark's Fifth Army, excluding the force at Anzio, had six divisions in line including the four that made up the French Expeditionary Corps. At Anzio, VI Corps, part of Fifth Army and under Truscott's command, had been reinforced to number 150,000 men. Truscott's VI Corps included six infantry divisions (the British 1st and 56th Divisions, the US 3rd, 34th, 36th and 45th Divisions), Harmon's 1st Armored Division of 232 tanks and a Canadian-American First Special Service Force brigade. More than one million tons of materiel had been stockpiled at Anzio ready for Operation Diadem. The key concept of the operation was the breakout from Anzio. It needed to capture the Alban Hills then cut Highway 6 at the mountain town of Valmontone. If VI Corps could block Highway 6, the bulk of the German Tenth Army would be trapped. On the day Operation Diadem was launched, 11 May 1944, Lucian Truscott wrote to Sarah: 'We are on the eve of great things. I hope this summer will carry us a long way toward the end of the war in Europe.'

Four plans of action

To contribute to those 'great things' about to happen, Truscott had devised four possible plans of action for the Anzio breakout, each with its own codename. Turtle and Buffalo were the primary ones with two others, Crawdad and Grasshopper, considered remote possibilities. Operation Turtle had Rome as its ultimate objective, moving through Aprilia, Campoleone and Albano then driving directly down Highway 7 north-west to Rome. Buffalo attacked in a different direction north-east to Cisterna across Highway 7. It then continued east of the Alban Hills, pushing on to Valmontone and cutting Highway 6. This was exactly what Harding had envisaged as VI Corps's part of Operation Diadem. Highway 6 was not only the German's main supply line for their forces around Cassino but it would also be their main line of retreat to Rome.

When General Alexander arrived at VI Corps Headquarters at Nettuno on 5 May, Truscott briefed him on the four options his staff had devised. While Clark favoured the Turtle option, Alexander clearly did not. He stressed to Truscott that Buffalo was the only plan to follow, and that evening issued an order that Rick Atkinson describes as one 'of uncommon clarity'. It directed Clark's Fifth Army 'to cut Highway 6 in the Valmontone area, and thereby prevent the supply or withdrawal of troops from the German Tenth Army'. Clark was furious, accusing Alexander of 'trying to run my army' and believing that the British were trying to deny him the glory of capturing Rome. He told Truscott the next day that the 'capture of Rome is the only important objective'. It was to remain Clark's principal intent. While Clark assured Alexander that his Fifth Army would do its part in Operation Diadem, he had no intention of implementing the Buffalo plan. He ordered Truscott to keep all four options up to date.

It took longer than expected for the overwhelming force of Operation Diadem to break through the Gustav Line. By 17 May, after five long months and another week of bitter fighting, the Germans withdrew rather than risk encirclement. The soldiers from two Allied armies advanced north in pursuit. The time was rapidly approaching for the expanded VI Corps to join the fight.

The breakout

On 20 May Truscott received the order from Mark Clark to launch Operation Buffalo. D-Day for the attack was at 6.30 on the morning of 23 May. For weeks prior to this new attack VI Corps had been firing early-morning artillery barrages. The Germans became used to it as part of their daily routine. As Truscott recalled of the early morning of 23 May, there was 'no sight or sound to indicate that more than 150,000 men were tensely alert and waiting'. This soon changed. At precisely 5.45 a.m. more than 1000 guns joined the barrage as did the fire from tanks, tank destroyers and mortars. A barrage from hundreds of machine guns joined in. It halted at 6.25 a.m. when 60 fighter-bombers from the Army Air Force attacked the German positions. When they had finished, the barrage resumed as the infantry and tanks of VI Corps surged towards the German positions.

Truscott had planned for four major attacks and two minor ones to open Operation Buffalo. In the centre of the Anzio position the 1st Armored and 34th Divisions were to attack due north between Cisterna and Velletri to cut Highway 7. As they did so, 3rd Division on their right flank would attack Cisterna and encircle it. On their right flank, the 1st Special Service Force, roughly the size of a regiment, was to attack along the Mussolini Canal towards Highway 7, protecting the right flank of the breakout. On the left of 34th Division, the 45th Division would expand the breakout to the west, attacking towards Carano. The newly arrived 36th Division was in reserve behind the 3rd Division and would exploit any success the dogfaced soldiers achieved. On the west flank of the beachhead were the British 1st and 5th Divisions, both at around half of their normal strength. Truscott did not plan to use them in the breakout but directed them to make limited attacks in order to hold some of the German defenders in place.

The fighting was intense as the Germans resolutely defended their positions. By the end of 23 May, most of VI Corps units had taken their objectives set for the first day. The 3rd Division was just 600 yards from Cisterna, having advanced more than a mile. It had also suffered

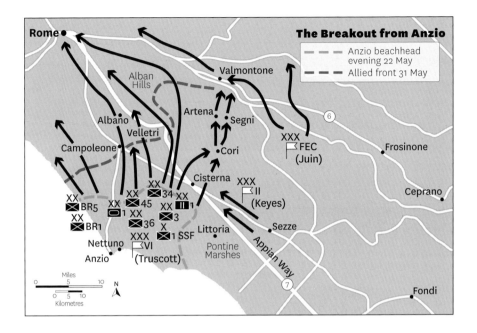

over 1000 casualties, 3rd Division's most costly day during the war and among the worst casualty figures for any infantry division during the war. On the right, the 1st Special Service Force had crossed Highway 7 below Cisterna only to be driven back by German Tiger tanks. On the far left, the British divisions, making the diversionary attacks, captured no ground, while in the centre the 45th Division attack gained little ground and they were counterattacked by German Tigers. The division suffered 458 casualties.

Between the British Divisions and the 45th Division, Ernie Harmon's 1st Armored Division attacked along a three-mile front with 232 tanks. He expected to lose heavily, perhaps as many as 100 tanks in the first hour. The attack was not helped when Combat Command B on the Division's right wing, barely a quarter mile from its start line, blundered into a poorly marked Allied minefield, losing 30 tanks. Combat Command A on the Division's left made much better progress, capturing some important high ground and bringing Highway 7 within gun range. The day cost Harmon's 1st Armored Division some 86 tanks and tank destroyers.

Truscott takes the opportunity to write a quick letter, on 20 May 1944. Three days later, VI Corps, now swollen to 150,000 men and the size of an army, launched the breakout from the Anzio beachhead.

Losses in tanks and guns could be quickly replaced. Losses in men could not be so easily overcome. Fifth Army casualties for this day were more than 2000, with half of these from the 3rd Division. Rick Atkinson noted that this was 'the highest single-day tally in the Italian campaign'. It was equally savage for the Germans, with some 1500 prisoners taken and some units losing over half their strength. It was an old-fashioned battle of attrition which the Germans had to lose if it lasted much longer. Cisterna was now threatened with envelopment.

The 24 May saw more heavy fighting and several German counterattacks that were defeated only with close air support. Truscott was satisfied with the progress being made, writing that: 'It was clear we were through the main German defences.' It was evident that German resistance was weakening. Harmon's tanks had made considerable progress, looping behind Cisterna from the west, while the 3rd Division advanced east of the town. By nightfall both divisions had nearly surrounded the town and a further 850 prisoners were taken. Highway 7 had been cut north and south of Cisterna and a reconnaissance force had pushed north to be just four miles from Velletri on the lip of the Alban Hills.

Truscott was somewhat alarmed, though, when Lieutenant General Clark arrived at his headquarters on 24 May and asked if he had considered shifting his attack to the north-west and Rome. Truscott replied he would do so if the Germans reinforced Valmontone and it proved impossible to take the town and cut Highway 6. Clark seemed satisfied and left but wanted to be continually updated on progress being made. Truscott was concerned enough to make a handwritten amendment to the diary kept by one of his aides, something he rarely did: 'Clark asked if I had thought of possibility of changing dir att [direction of the attack] to NW. Had staff working on one — only to be used in case Germans moved H.G. [Higher ground] and Para Div moved to Airborne Gap.'

On 25 May the 3rd Division captured Cisterna after three days of heavy fighting. Truscott's battle casualties were now more than 2500. On that day units from VI Corps linked up with a Fifth Army corps advancing from the Gustav Line. This was II Corps, commanded by Geoffrey Keyes, and it had advanced from the Cassino front, covering 60 miles in two weeks. The Anzio beachhead was no longer isolated. When he learned of the link-up, Mark Clark raced to the scene and ordered it to be restaged for the cameras and the two dozen war correspondents who were part of his entourage. In his mind, the battle of Anzio was over and the battle to take Rome was on. After 125 days Anzio was no longer an isolated beachhead behind enemy lines.

Clark's change of plan

Truscott was, in his own words, 'rather jubilant' about the success of the breakout from Anzio. He found time to write to his 'Beloved Wife', Sarah:

> I know that you have been following events here by radio and papers. You will know that we have at last gotten underway — and I might add — our lads have really been going to town. Needless to say we are as happy as can be to have our period of inactivity . . . finally ended and we have a chance to take a few good licks at the Boche. Believe me our country can really be proud of these soldiers . . . they're really doing one grand job.

On 25 May, the day of the link-up with the rest of Fifth Army, his old 3rd Division was clearing Cisterna, house by house, other units of VI Corps had reached Cori six miles to Cisterna's north-east and over 2500 prisoners had been taken. Harmon's 1st Armored Division continued to advance north against heavy opposition but making steady progress. At the point of the spear, a four-battalion force led by Colonel Hamilton Howze, advanced along a gentle valley to the east of the Alban Hills. Soon Howze had a dozen Sherman tanks within half a mile of Highway 6. He reported that: 'I am in a soft spot. For Pete's sake, let the whole of 1st Armored come this way.' As Truscott arranged to move the bulk of VI Corps' firepower to Valmontone to entrap the retreating Germans, he returned to his headquarters after a visit to the front line to find that the axe had fallen on Operation Buffalo.

Awaiting Truscott at his headquarters with his sombre-looking VI

Always keeping an eye on the ball, Truscott checks on the status of his VI Corps against the positions marked on a map in his headquarters.

Corps staff was Brigadier General Donald Brann, Clark's operations officer. Brann wasted no time: 'The boss wants you to leave the 3rd Infantry Division and the Special Force to block Highway 6, and mount that assault you discussed with him to the north-west as fast as you can.' It was a decision that would have catastrophic consequences. Truscott was being ordered to abandon Operation Buffalo, just as its success looked likely, to replace it with Operation Turtle. While Turtle was the shortest route to Rome, it entailed a frontal assault on a German defensive position now reinforced with the 1st Parachute Corps, the toughest soldiers in the German army. It also required that Truscott pivot the direction of his attack by 90 degrees to the west, leaving just the understrength and battered 3rd Division and Special Force to push on to Valmontone. As he recorded: 'A more complicated plan would be difficult to conceive.'

The change in plan made no sense at all and Truscott protested vigorously. He asked Brann if he could discuss this change with Clark, only to be told that his superior could not be contacted by either radio or phone. Brann was emphatic that Clark wanted the change made. He was blunt: 'There's no point in arguing. It's an order.' Truscott felt he had no choice but to obey the order of his army commander. Any hint that he lacked confidence in this change or was dissatisfied with it was, Truscott believed, 'poor leadership'. He directed his staff to prepare the plans that would swing the 45th, 34th, 36th and 1st Armored Divisions off in a new direction of attack against a skilled enemy in well-prepared defensive positions. Once again Truscott took the unusual step of adding a handwritten note to his aide's diary. It read that after meeting with Clark on the evening of 25 May he: 'Protested to Clark then.'

While Truscott loyally followed Clark's order to switch the direction of attack, he revealed his true feelings in his highly regarded war memoirs. He wrote that on receiving the order to switch the direction of the attack, he was 'dumbfounded. I protested that the conditions were not right.' The consequences he believed were dire. 'Such was the order that turned the main effort of the beachhead forces from the Valmontone Gap and prevented the destruction of the German X Army.'

Colonel Hamilton Howze, at the tip of VI Corps' spear and gazing down at the undefended Highway 6, the road by which the German army planned to escape, for the rest of his life believed that Clark's decision to ignore Valmontone, cutting Highway 6, and to focus on capturing Rome was wrong. He told an interviewer in 1972 that it was 'one of the worst decisions I ever knew. At the time I thought so, and in retrospect, I still think so.' Watching the action from London, British Prime Minister Winston Churchill was alarmed to see a great opportunity slipping through the Allies' fingers at Valmontone. He sent a warning note to Alexander on 28 May: 'At this distance it seems much more important to cut their line of retreat than anything else. ... A cop is much more important than Rome, which would anyhow come as a consequence.'

While Alexander knew 'a cop' was what mattered and had planned to entrap the German X Army in Operation Diadem, Clark did not see it that way. It was Rome that mattered most to Clark, and he was determined that it would be his Fifth Army that captured it.

While the 3rd Division and 1st Special Force pushed on to Valmontone, the main force of Truscott's VI Corps pivoted to the north-west and began the advance on Rome. They immediately ran into strong German resistance exactly where Truscott expected it. Blocking the Turtle route were the German 3rd Panzer Grenadier Division, the 65th Division and the 4th Parachute Division dug in along the western face of the Alban Hills in a new defensive position they called the Caesar Line. Four of Truscott's divisions attacked these German defenders for two days, making little progress and taking heavy casualties. Harmon's 1st Armored Division had suffered particularly heavily, so much so that Harmon believed by 30 August his division could fight for only one more day.

Then on 30 August Truscott received the break he needed. Major General Fred Walker's 36th Division was committed to the fight. This formation had a reputation for being a 'hard luck' division and had suffered heavy casualties in the Cassino battles. Walker was determined to avoid the unnecessary casualties that a frontal attack would entail, and he made a thorough examination of the battlefield before committing his division to action. This involved studying old maps of the Alban Hills, conducting an aerial reconnaissance in a light plane, sending out scouts and walking over the ground. Walker had been a mining engineer and he believed he had found a way to outflank the Caesar Line. This was an old logging road that reached the top of the Alban Hills near Velletri on Highway 7. Better still, the area at the top of the road was undefended with a two-mile gap between the left flank held the German 1 Parachute Corps and the right flank of the panzer divisions. At first Truscott was sceptical about making an attack here, especially with just an infantry division. But Walker was persistent, backed up by his division's engineer who assessed that the volcanic nature of the soil would enable the construction of a military road. Truscott thought it over and admitted to Walker: 'You may have something here.' Although his own VI Corps engineer opposed the scheme, Truscott could see no alterative except another costly frontal assault. He decided to take the risk. In typical Truscott style, he sent Walker an entire engineer regiment to assist and warned: 'And you had better get through.'

They did. As one of Walker's regiments pinned the German defenders with a frontal assault on Velletri, two other regiments climbed the old logging road to get behind the town. By midday on 30 May, some 6000 American soldiers were spread along the southern lip of the Alban Hills behind the German Caesar Line. Hard on their heels were fifteen engineers' bulldozers grading a road wide enough for Sherman tanks and tank destroyers. Any obstacle too big for bulldozers to handle were demolished by the sappers. It took a night and a day for the road to be completed but immediately after it was finished the new road was filled with tanks, self-propelled guns, signallers with field telephones and artillery observers. The Germans at first did not think the Allies had penetrated their line in such strength. When a counterattack was ordered by Kesselring on the evening of 31 May to seal the gap, the salient behind Velletri was five miles deep and held by thousands of American soldiers.

By 1 June Velletri was completely surrounded and American soldiers had advanced to the highest peaks on the Alban Hills. Only eleven

36th Division soldiers had been killed securing the heights. More were killed, though, in the fight to take Velletri that day. At Valmontone, 85th Division of II Corps linked up with the 3rd Division and began the push down Highway 6.

The road to Rome

Mark Clark now had eleven divisions advancing north and converging on Rome. But they were moving much too slowly for his liking. On 2 June his directive to Truscott was: 'The attack must be pushed to its limit.' He wrote in his diary that day: 'This is a race against time, with my subordinates failing to realise how close the decision will be.' He did not have much longer to wait. Valmontone was abandoned by the Germans on the same day Clark was complaining about his subordinates' lack of concern for his reputation. A reconnaissance along Highway 6 by a battalion from 3rd Division did not make contact with the enemy. By dawn on 3 June both Truscott's VI Corps and Keyes' II Corps had a clear road into Rome. Truscott found time on 3 June to write to Sarah outlining the good news:

> I have not written for four or five days — but as you can well imagine I have been a busy man these last few days. However, it is paying dividends — and the last lap of the road to the Eternal City is beginning. All of this will be old when you read these lines — for you will have heard by radio or read by paper all the news there is about these doings. Everyone is tired of course but every one realizes that the 'other guy' is far worse off than we. I'm really proud of our lads and their accomplishments during this time.

That evening his aide's diary recorded a significant event: 'Berlin broadcast announces evacuation of Rome. General [Truscott] confident of breakthrough tomorrow.'

On 4 June units from 36th, 85th and 1st Armored Division met on the outskirts of Rome. There was some argument over which formation should enter Rome first before Truscott arrived and directed 1st Armored Division into Rome to secure the bridges over the Tiber River. It was a symbolic moment. The Allies had finally liberated the first European capital from Nazi oppression.

Mark Clark did not enter Rome until 5 June. After getting lost with his entourage in Rome's winding streets, Clark eventually found his way to the town hall where he had summoned his corps commanders to meet him. Clark then gave a brief victory oration to the assembled reporters, proclaiming that: 'It was a great day for the Fifth Army.' As one reporter there caustically commented:

> That was the immortal remark of Rome's modern-day conqueror. It was not, apparently, a great day for the world, for the Allies, for all the suffering people who had desperately looked toward the time of peace. It was a great day for the Fifth Army. (Men of the Eighth Army, whose sector did not happen to include Rome but without whose efforts this day could not have occurred, did not soon forget the remark.)

Clark's ungracious speech left his corps commanders 'red with embarrassment', while many of the reporters there were also discomforted, with one stating 'On this historic occasion I feel like vomiting.' Truscott was disgusted, writing later that 'I was anxious to get out of this posturing and on with the business of war'.

Farewell to Italy and Mark Clark

Truscott was soon back at the front leading VI Corps 85 miles north of Rome and capturing two of the heavy railway guns that had made the

Anzio beachhead such a hazardous place to be. He wrote to Sarah on 11 June outlining the success of VI Corps and stressing what lay ahead for the soldiers now in France after the Normandy landings:

> ... we have been pressing hard and fast on the heels of the fleeing Krauts as you will know from reading the papers and listening to the radio. It has been a successful campaign as you know — and my lads have covered themselves with distinction. Right now we are further advanced than anyone else. I'm certainly proud of them all.
>
> I suppose the whole country is agog over news of the invasion [as] we certainly are and I'm sure everyone is praying for them. God knows they need all the help they can get — having experienced all major landings over here except that one — I speak with deep feeling on that score.

As he paused for the next assignment, Truscott learned that he and part of his VI Corps would soon be leaving Italy and heading to France.

There is little doubt that Truscott had performed well at Anzio despite his ill health. Rome's capture owed much to his success. Truscott had taken over VI Corps at its lowest ebb, but he quickly restored its morale and aggressive spirit. When the time came for the breakout from Anzio, Truscott's VI Corps had achieved all its objectives and the capture of Valmontone, entrapping the German X Army south of the town, was within its grasp. Ordered to change the direction of the attack, against his instincts and wishes, Truscott loyally did so and eventually found a way to break through the German Caesar Line, opening the road to Rome. It was a bitter triumph for Truscott, and he regretted Clark's decision to switch the direction of his attack for the rest of his life. As he wrote in his account of the war:

> There has never been any doubt in my mind that had General Clark held loyally to General Alexander's instructions, had he not

changed the direction of my attack to the northwest on May 26th, the strategic objective of Anzio would have been accomplished in full. To be first in Rome was poor compensation for this lost opportunity.

In many ways, Anzio was Truscott's finest hour and marked him as a commander whose star was on the rise. But Anzio exacted a heavy price for Truscott. His son, Lucian III, wrote that Anzio was 'apparently the worst of all his experiences'. Tears would come into his eyes when he described how many young men on both sides had been killed in the beachhead. The sheer scale of the death and destruction there overwhelmed Truscott at times. As he said to his son, 'There would always be something of him on that damned beachhead.'

As noted earlier, Truscott and Clark had a troubled relationship. Clark wrote in his diary on 8 March that Truscott was 'a difficult subordinate to handle. He makes demands, knowing full well that many of them can't be granted.' Clark's diary for the months Truscott was in command of VI Corps made repeated references to Truscott's 'whining attitude'. For his part, Truscott twice reminded Clark that he could relieve him of the command of VI Corps 'at any time'. When the time came for Clark to formally assess Truscott's performance in his officer efficiency report, he did not rate Truscott that highly: 7 out of 20 and then 20 out of 250.

One man who knew the truth about Truscott's abilities was Lieutenant General Jacob Devers, the senior American commander in the Mediterranean. Devers had no time for Mark Clark and refused to deal directly with him. It was Devers who picked Truscott's next assignment. Devers believed that Truscott was 'the finest army combat commander in the battlefield today' and selected him to lead VI Corps for the forthcoming invasion of southern France. He made sure to allocate Truscott two of the best divisions then serving in Italy. Truscott now needed to prepare for his fifth amphibious landing of the war and to serve under a different army commander. As he made ready for this new assignment, Truscott had no idea that he would have to return to Italy before the end of the year and serve again under Lieutenant General Mark Clark.

6 Fighting in France, August to October 1944

A new army and army commander

After the capture of Rome in early June 1944, VI Corps now pared down to just three divisions under Truscott, prepared for a new mission in southern France. Also leaving Italy for France were the French Corps under General Alphonse Juin and the 2nd British Parachute Brigade. Together these formations formed the reactivated Seventh Army. This army had been resurrected for the purpose of undertaking Operation Dragoon, an amphibious landing on France's southern coast.

Truscott's VI Corps was designated as the initial assault force for Operation Dragoon, opening the door for the Allies into southern France. While the VI Corps now consisted of just three divisions instead of the seven that had been at Anzio, Truscott was allowed to choose which divisions he could take to France. It was no surprise that Truscott's first choice was the 'Rock of the Marne', his beloved 3rd Division which had been a Regular Army formation. The other two divisions Truscott chose were the 36th, a division formed by the Texas National Guard, and the 45th Division, made up of National Guard units from Arizona, New Mexico and Colorado. Truscott selected the 3rd and 45th Divisions because of their training and experience in previous amphibious operations. The 36th was chosen because of its outstanding performance during the Anzio breakout. Truscott's protégé, Major General 'Iron Mike' O'Daniel, still commanded the 3rd Division, while the 45th Division was commanded by the equally impressive William W. Eagles who had served under Truscott at Anzio. The 36th Division had a new commander. Replacing Major General Fred Walker was Major General John E. Dahlquist, an officer who had worked mainly in logistics, staff and training roles. His appointment to the 36th Division was Dahlquist's first combat command and Truscott would soon have concerns whether Dahlquist was up to the task. There was no doubting Truscott's abilities, though. It had been clear to Jacob Devers and many others that Truscott was the ideal man to plan and lead the initial invasion force for Operation Dragoon. As Russell Weigley has written: 'It would have been hard to find a better commander for the amphibious assault. . . . Truscott combined a horse soldier's dash with a wealth of amphibious experience.'

Commanding the Seventh Army and Truscott's immediate superior officer was a man he had never met: Lieutenant General Alexander M. 'Sandy' Patch. Patch, much to Truscott's relief, was nothing like Mark Clark. While his father had served in the cavalry, Patch opted for the infantry instead. Patch had been sent to the Pacific in 1942 where he had organized the defences of New Caledonia before leading an army division to relieve the Marines on Guadalcanal and finally cleared the islands of Japanese. He had a solid reputation for quiet efficiency and courage. Truscott wrote of their first meeting:

> I was fully aware of his reputation and knew he was highly esteemed. He was thin and wiry, simple in dress and forthright in manner — obviously keenly intelligent with a dry Scottish humor. His quick and almost jerky speech and movement gave me the impression he was nervous and found some difficulty in expressing himself. . . . I came to regard him highly as a man of outstanding integrity, a courageous and competent leader, and an unselfish comrade-in-arms.

While the two would not agree on everything, their working relationship was good and was certainly a lot smoother than the one Truscott had had with Mark Clark in Italy. For Lieutenant General

'Sandy' Patch, working with Truscott's VI Corps was much smoother than working with the French corps that formed the other part of his Seventh Army.

Planning Operation Dragoon

The plan to invade southern France had a long gestation and was not without controversy. Initially called Operation Anvil, the invasion of France's southern coast was to have occurred in June 1944 in conjunction with Operation Overlord, the cross-channel invasion of the Normandy beaches. In April 1944, General 'Ike' Eisenhower had been forced to cancel the southern France operation. There were simply not enough landing craft available to do two concurrent amphibious landings. However, by late June 1944, with Rome finally taken and the invasion of northern France bogged down in the *bocage* (mixed woodland, pasture and hedgerows) country, and with the landing craft now free to use, a second amphibious landing in France was now a viable option.

British Prime Minister Winston Churchill and the British service chiefs vigorously opposed this new landing, arguing instead that further resources should be committed to Italy and the Balkans. Eisenhower, however, stuck to the view that his forces in France needed the major ports of Marseille and Toulon to sustain the Allies in Europe and that the invasion of southern France was therefore a military necessity. He later recalled that difference of opinion 'brought up one of the longest-sustained arguments that I had with Prime Minister Churchill throughout the period of the war.' Churchill lost the argument and when it was necessary to change the name of the operation for security reasons, he insisted that it now be codenamed Dragoon to show his unwillingness about a venture that he had been 'dragooned' into. Truscott was delighted with the change of name, later writing that 'as an old cavalryman, I took this change of name to be of good omen'.

The plan for Operation Dragoon envisaged the capture of the ports of Marseille and Toulon followed by an advance north up the Rhône River valley, liberating the cities of Lyon and Dijon before linking up with the Allied forces advancing across northern France from Normandy. For political reasons, the capture of the two main ports of Marseille and Toulon were to be undertaken by French troops. The role allocated to Truscott's VI Corps was to land first to the east of the two ports, secure a beachhead and push inland to seize the mass of hills that dominated the area. The next day, with VI Corps protecting their right flank, the French II Corps would land and advance on Toulon and Marseille. Truscott liked the plan but stressed to Lieutenant General Patch that he needed to set up an invasion-planning headquarters with representatives from the army, navy and air force at Naples as soon as possible. Truscott also emphasized the need for comprehensive training, including rehearsals, and to establish an undivided command responsibility during the initial stages of the landing. Patch fully concurred with Truscott's suggestions.

For a month prior to the invasion, Allied bombers pounded every potential target on the southern coast. Then, just days before the invasion day, the bombers focused on Marseille in an attempt to deceive the Germans that this was where the invasion would come. Truscott's VI Corps would lead Operation Dragoon set for 15 August 1944. It would commence on the night of 14 August with the 1st Special Forces Regiment capturing two offshore islands and landing at Cape Nègre in order to destroy the German coastal artillery batteries located there. Then at 4 a.m. on D-Day, 15 August, American and British airborne troops would drop onto the rugged hills near Le Muy and attack the German defences from the rear. After an intense naval and air bombardment of the various landing beaches, Truscott's three divisions were to land on the French coast along 30 miles of the shoreline. On the left flank, near Saint-Tropez, was the 3rd Division's landing zone. In the centre, near Sainte-Maxime, was the 45th Division's zone with the 36th Division landing on the right near

Fréjus. Once the beachhead was secure, a French armoured brigade was to land to the left of VI Corps. The French divisions planned to land on 16 August with all of Seventh Army ashore by D+3. Then the entire invasion force would converge on Toulon and Marseille. Once the ports were captured, the Allied forces planned to advance north up the Rhône Valley to link up with the Allied armies in the north. Facing them in southern France was the German Nineteenth Army, headquartered at Avignon, and consisting of eleven divisions, including two panzer divisions, and various odd units of artillery, armour and coastal guns.

Prior to the invasion Truscott insisted on implementing a detailed training programme which included a full dress rehearsal for all three divisions of VI Corps. He was pleased with the progress made, later noting that: 'Few divisions have ever been better prepared for the task which lay ahead.' The tanks for VI Corps would be provided by a French armoured brigade of the French 1st Armoured Division. The French commander, Jean de Lattre de Tassigny, was not happy with this arrangement and berated Truscott at a lunch for inspecting the formation without his permission. Truscott referred the matter to Patch who confirmed that Truscott had command of the French armoured brigade during the initial landing. Truscott knew that this arrangement could not last once the Allies started their advance up the Rhône Valley, so he created his own armoured force. Placing Brigadier Fred Butler, his assistant corps commander, in charge of it, Butler Force would consist of some tanks and tank destroyers, engineers, and motorized infantry with staff and communications provided by VI Corps Headquarters. While it would always be an ad hoc force and considerably weaker than an armoured brigade, Butler Force did provide Truscott with an armoured force should he need it.

Nine days before Operation Dragoon, his fourth time in charge of an invasion force, Truscott spoke to the officers in VI Corps. His speech to them indicated much about Truscott's command style. He told his officers that:

'Our American soldiers are, by nature, hunters. The vast proportion of these lads are farm-raised or have a farm background. They are all hunters by instinct. Encourage that hunting instinct. Make every soldier go into every fight feeling like a hunter. Finally, I say to you, get forward. No battle was ever won by any commander sitting in a dugout with a telephone in one hand and a radio in the other. You have to know from your own personal observation what the score is. I say to you, from Division Commander down to squad leader, get forward.'

As always, getting forward was what he intended to do during Operation Dragoon.

The last amphibious landing

The Operation Dragoon landings took place on 15 August along the French Riviera coastline between Cape Nègre in the west and Fréjus in the east. It was a large invasion force, bigger than that used on 6 June, consisting of some 885 ships, 1400 landing craft, 20,000 armoured and transport vehicles and 150,000 men. Truscott's VI Corps with 95,000 men and 1100 vehicles led the way. Sailing on the USS *Catoctin* from Italy, Truscott was pleased when Lieutenant General Patch told him that he expected Truscott to run his own battle once ashore. It was exactly what Truscott wanted.

All accounts of the landing on 15 August stress how smoothly it went. US soldier cartoonist Bill Mauldin later described it as 'the best invasion I ever attended'. Standing on the railing of the *Catoctin*, Truscott watched as the infantry from his three divisions headed to their designated beaches. All proceeded according to plan except for a regiment of the 36th Division which had been deliberately landed at an alternative beach after the navy commander deemed the original beach too dangerous to land. This was the one glitch in an otherwise

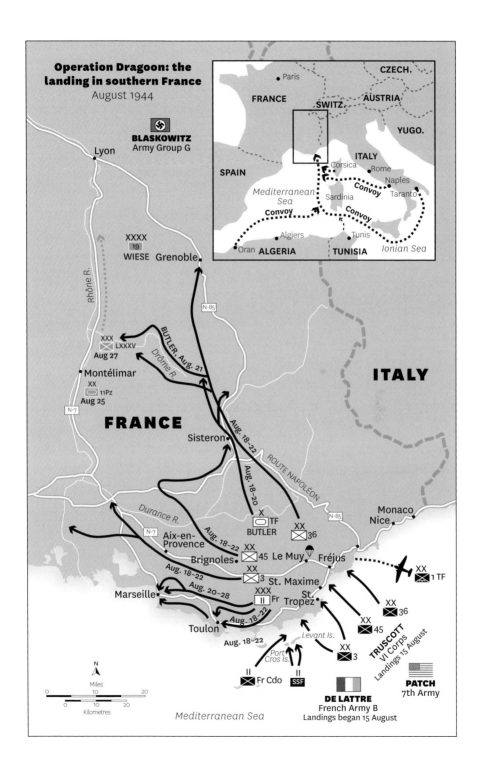

perfect amphibious landing. It was not a great start for Major General Dahlquist under Truscott's command.

Truscott went ashore that morning accompanied by a French liaison officer who had been at Port Lyautey in North Africa with the Vichy French. By late afternoon he had a good picture of how his VI Corps had performed, and he was happy with the progress made. Both the 3rd and 45th Divisions were already on their final objective and 36th Division looked like securing theirs later that morning. Truscott was especially pleased at the low number of casualties, just 183 across VI Corps. German casualties were substantially higher and on this first day over 2000 German soldiers had been taken prisoner. Truscott later reflected that 'speed and power, thorough planning, and training, had paid off'.

Truscott's landing on 15 August was followed up the next day with three more divisions from French II Corps landing and advancing on the two key ports on the south coast. Progress continued to be easy. By 18 August the Allies had advanced more than 50 miles from the coast and the major ports of Toulon and Marseille were surrounded. They held out until 28 August, captured a month ahead of schedule, and giving Eisenhower the large ports he desperately needed. Over the next month Operation Dragoon grew to include two complete Allied armies: the French First Army (under Jean de Lattre de Tassigny) and US Seventh Army (Patch) eventually coming together as the Sixth Army Group commanded by Lieutenant General Jacob Devers.

The German Nineteenth Army did not stay to fight it out on France's southern coast. On 16 August, orders from Berlin instructed the German defenders to stage a fighting withdrawal northwards up the Rhône Valley and into Germany. Excluded from this withdrawal order were the units defending Marseille and Toulon who were expected to defend the ports until the end.

Advancing up the Rhône Valley

Truscott was happy with the way the landing had progressed. But he wanted more. Knowing that the critical point of an offensive action was 'to destroy the enemy', he sought permission from Patch to activate Butler Force and send it deep behind enemy lines. General Patch readily approved Truscott's concept of operations and Truscott readied Butler Force at the town of Le Muy. Butler Force now consisted of a motorized infantry battalion, two medium tank companies, a tank destroyer company, a light cavalry squadron, and one self-propelled artillery battalion. For Truscott, the key was for Butler Force to reach the town of Montélimar on the Rhône River's right bank and entrap the Germans south of the town. Butler Force set off on the morning of 18 August, heading into the mountains in the general direction of Grenoble.

Butler Force made excellent progress and in a single day covered almost 50 miles, reaching Sisteron and taking more than 1000 Germans prisoner. The town of Sisteron was about a third of the way to Grenoble and almost halfway to Montélimar. Truscott was impressed and immediately ordered the 36th Division to follow Butler's route and meet up with him in Sisteron. Butler was told to wait at Sisteron until the 36th Division arrived but to scout ahead to determine how best to seize the high ground north of Montélimar. With German resistance weak everywhere, Butler's scouting force advanced a further 45 miles on 19 August and covered a similar distance on 20 August.

Knowing that time was critical, Truscott changed his orders on the night of 20 August. It was time to shut the gate on the Germans south of Montélimar. In an emergency message carried by a colonel in a jeep, Truscott directed Butler to immediately shift his entire force to Montélimar without waiting for the 36th Division. Butler did not need prompting. On 22 August he had his force on the heights above the town, having covered 200 miles in four days. Truscott was happy with this progress but was soon to be disappointed. On the afternoon of 21 August, intelligence from Ultra intercepts revealed that the 11th Panzer

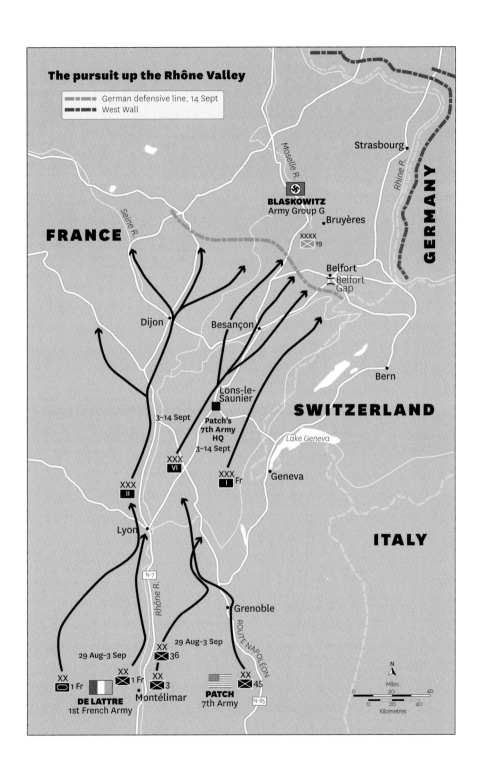

Division had crossed the Rhône and was heading for Aix-en-Provence, 30 miles north of Marseille. This information caused Truscott to slow down the advance of 3rd and 45th Divisions and to keep back one regiment of the 36th Division. It proved to be an unnecessary delay as a hasty counterattack by 11 Panzer Division against the 3rd Division at Brignoles was easily beaten off.

On 24 August, alarmed that his trap had still not been sprung, Truscott flew with an aide by Cub aeroplane to Aspres-sur-Buëch which was about halfway between Sisteron and Montélimar. He had arranged to meet with Dahlquist to make sure that his subordinate understood his role in the plan. Truscott was annoyed that Dahlquist was not at Aspres-sur-Buëch to meet him, having sent his chief of staff instead. His annoyance turned to anger when he learned that Dahlquist did not have any units moving on Montélimar, having directed them to prepare for German counterattacks emerging from Grenoble and Gap. Truscott immediately arranged for a regiment from 45th Division to contain any threat from Grenoble and directed, yet again, that the bulk of 36th Division move on Montélimar at once. He sent a terse note to Dahlquist: 'Apparently, I failed to make your mission clear to you. The primary mission of the 36th Infantry Division is to block the Rhone Valley in the gap immediately north of Montelimar.'

While Butler informed Truscott that he had captured more high ground north of Montélimar, aerial reconnaissance confirmed Truscott's worst fears. The Germans were still escaping on both sides of the Rhône. Butler did have enough units to capture all of the high ground around Montélimar and the Germans were finding sufficient gaps to squeeze through. To make matters worse, Dahlquist had arrived at Montélimar and disbanded Butler Force, absorbing it into his own division. Truscott immediately countermanded this decision. He pointed out to both Dahlquist and Butler what they had done wrong so far, reminding them that 'the essential task' was to deny the Germans the highway north of Montélimar by holding the long ridgeline there.

The day after receiving this directive, Truscott was incensed that the long ridge overlooking the highway north of Montélimar was still not captured and that Germans were escaping around the town. He moved to the town himself and to 36th Division's Headquarters in a mood reminiscent of Patton's tantrum in Sicily. Truscott, however, dealt with his frustrating subordinate in an entirely different manner to Patton. Meeting with Dahlquist, Truscott bluntly stated:

> 'John, I have come here with the full intention of relieving you from your command. You have reported to me that you held the high ground north of Montelimar in that you had blocked Highway 7. You have not done so. You have failed to carry out my orders. You have just 5 minutes in which to convince me that you are not at fault.'

Dahlquist, his military career in the balance, outlined the reasons for his failure to implement Truscott's orders. According to Dahlquist, some of his subordinates had misinformed him as to where their units were located. He had corrected that mistake and he now had four battalions of artillery covering the highway which he thought was enough to stop enemy traffic. Apart from the initial mistake, Dahlquist asserted that he had done all that he could to follow Truscott's intent. Truscott recorded his change in mood: 'I did not fully concur, but I decided against relieving him.'

With his block now in place, on 25 August the 3rd Division pursued the Germans north to Montélimar, which they reached on 27 August. Many German vehicles, guns, horses and men were destroyed by the guns of VI Corps. Truscott recorded of it: 'I know of no place where more damage was inflicted upon troops in the field. And the sight and smell of the section is an experience I have no wish to repeat.' While not attaining the complete success he wanted, Truscott's ambush at Montélimar had been devasting for the Germans, resulting in the destruction of 4000 vehicles and the capture of 5000 prisoners. In the ten days since the landing, his VI Corps had advanced just over 100 miles, taken 23,000 prisoners and destroyed two complete German divisions. Its own casualties numbered 1300. Truscott later reflected on this period: 'Even if Montelimar had not been a perfect battle, we could still view the record with some degree of satisfaction.' That it did not achieve more was not Truscott's fault. He had done everything he could to achieve a decisive result, but the reality was that his blocking force was not strong enough, being more of a light screen than a fully organized fighting force. According to Rick Atkinson, 'Task force Butler had been too weak, the 36th Division too slow, the 3rd Division too cautious.'

Truscott's divisions continued to advance north-east up the Rhône Valley, meeting very little German resistance. In early September a reconnaissance by 36th Division indicated that the Germans had abandoned the city of Lyon. For political reasons it was left to the French to capture the city. After Lyon the French forces advanced north-west in order to link up with Patton's Third Army advancing eastwards across northern France, while Truscott's VI Corps headed for the Belfort Gap in an attempt to entrap more German forces.

Using the Doubs River as a major obstacle, the Germans cobbled together as many units as they could to make a stand at the city of Besançon. If the Germans could hold Besançon, many German units would escape from France through the Belfort Gap into Germany. Truscott was determined to prevent this. He advanced on the city with his three divisions abreast and on 6 September directed his best divisional commander, 'Iron Mike' O'Daniel, to take the city. He gave O'Daniel the option of either attacking Besançon with 3rd Division or holding the surrounding heights and awaiting reinforcements. 'Iron Mike' attacked the city and captured it two days later, taking over 700 prisoners.

Truscott's guiding principles

On the day Besançon was captured, Truscott issued an Operational Order to VI Corps that encapsulated his guiding principles. Issued on 8 September 1944 and probably meant more for the commanders of 45th and 36th Divisions, it read:

> The purpose of this operation is to destroy by killing or capturing the maximum number of enemy formations.
> Therefore, the following should be observed:
> a. Make every effort to entrap enemy formations regardless of size. Long range fires, especially artillery, will merely warn and cause a change in direction.
> b. All units, but especially battalions and lower units, must be kept well in hand. Commanders of all ranks must avoid wide dispersion and consequent lack of control.
> c. Tanks must accompany leading infantry elements and tank destroyers must accompany leading tanks. All must be supported by artillery emplaced well forward.
> d. Reconnaissance must be continuous and thorough — foot elements to a distance of five miles, motor elements to contact the enemy.
> e. Contact once gained must be maintained. The enemy must not be allowed to escape.
> f. Every attack must be pressed with the utmost vigor. Be vicious. Seek to kill and destroy.

With Besançon taken, the next obvious target for Truscott's VI Corps was Belfort. Lieutenant General Patch stressed to Truscott that 'The Belfort Gap is the Gateway to Germany'. Truscott did not need any urging. He knew the strategic importance of this location, situated between the Vosges and Jura mountains and between the Rhine and Rhône rivers. He planned to wheel his divisions to the right, approaching the city from the west. Each division was to keep one regiment in reserve.

Three United States three-star generals. On the left is the newly promoted Lucian Truscott, dressed in his trademark cavalry boots and pants and leather jacket, with cigarette in hand. In the centre is the Seventh Army commander Alexander 'Sandy' Patch. They are talking with Lieutenant General Jacob Devers, commander of the Sixth Army Group. All three men liked and respected each other, with Devers rating Truscott as the best of the US Army commanders. Patch tried to retain Truscott in his Seventh Army despite Truscott's elevated rank. It may explain why Patch does not look as happy as the other two lieutenant generals.

The attack on Belfort was delayed owing to fuel shortages and a change of priority in the Allied High Command that gave Field Marshal Bernard Montgomery's push into Holland top priority. Southern France, like Italy, was now to become a sideshow. Instead of attacking Belfort, on 15 October 1944 VI Corps attacked the town of Bruyères, about 100 miles further north and in the Vosges Mountains. After a tough fight, the town of Bruyères was captured on 19 October. It was Truscott's last battle in France. He was relieved of command of VI Corps on 24 October.

Relieved of command

The reason for Truscott's removal as VI Corps commander had nothing to do with his performance in France. Instead, it was the reward for his continued outstanding performance as a field commander. On 2 September 1944 the US Senate approved President Roosevelt's nomination to promote Truscott to Lieutenant General. Now a three-star general, Truscott was the same rank as his army commander Patch. On 17 October Truscott travelled to Épinal to meet with both Patch and Eisenhower. Eisenhower opened the conversation with news that shocked Truscott:

> 'Lucian, I am going to relieve you from VI Corps. You are an embarrassment to me now that you have been made a lieutenant general. All of my corps commanders now want to be made lieutenant generals. I'm going to assign you to organise the 15th Army.'

Truscott was not happy when it was explained that his new army command was an inactive one. It was a training and administrative role, with Truscott serving under General Omar Bradley. While Truscott liked and admired both Eisenhower and Bradley, he did not relish being confined to a desk away from the action. He asked

Eisenhower whether he could refuse the promotion, remain a major general and be kept in command of VI Corps. While Patch was in favour of Truscott remaining under his command, Eisenhower declined this request. Eisenhower had already found a replacement for Truscott, and he did not have the authority to revert Truscott in rank now that the Senate had approved his promotion. Truscott departed from VI Corps on 24 October 1944.

Earlier that month, on 2 October 1944, Truscott had appeared on the cover of *Life* magazine which ran an eight-page article on him. The cover photograph, taken by *Life* photographer George Silk, showed Truscott in his battle uniform of knee-high boots, cavalry britches, silk scarf around his neck, leather jacket and gleaming steel helmet. Truscott, square jaw jutting out, rested one foot on a rock while gazing in the distance as if searching for the enemy. The image captured many of Truscott's qualities. The article that accompanied another full-page image of Truscott, written by Will Lang and with background information provided by Sarah, outlined Truscott's achievements. It described Operation Dragoon as 'one of the most amazing exploits yet seen in amphibious warfare'. What Will Lang left out of his article was that Truscott had showed him a battle map and said, when he moved the pins on it a short distance, it meant that: 'By noon today 200 of my men will be dead.' It was typical of Truscott to outline the reality of war to a correspondent. Will Lang, while he never forgot the general's words and would later tell the story to Lucian Truscott IV, was too horrified to use it. Also, it was not what readers of *Life* wanted to hear. But Truscott knew only too well that when he issued orders men would be killed or injured no matter how well the battle ahead went.

After calling in to say farewell to Patch and the Seventh Army staff, Truscott did the same with General Jacob Devers and his Army Group staff. Next came reunions with Patton at Nancy and Omar Bradley at his 12th Army Group Headquarters in Luxembourg. On 30 October Truscott reported to General Walter Bedell Smith at the Supreme Headquarters Allied Expeditionary Force (SHAEF) in Versailles to begin the task of organising the 15th Army. Truscott was at SHAEF for less than a week. On 4 November Truscott and his staff departed for well-earned leave in the United States. He had been away for more than two years.

Truscott in France: an assessment

Truscott had achieved considerable success as commander of VI Corps in France. While Operation Dragoon had failed to entrap the German Nineteenth Army, it had severely mauled it. Of the 210,000 Germans who tried to evacuate southern France in August–October 1944, 80,000 had been lost in a month, most prisoners of war. Total casualties for the Germans numbered 140,000. This was a catastrophe for Nazi Germany, one that significantly contributed to their overall defeat. Capturing the parts of Toulon and Marseille so quickly and with little damage was also an important win for the Allies. With Antwerp, the second largest port in Europe, not being captured by the Allies until the first week of September 1944, the ports of Marseille and Toulon became major logistical hubs vital to the success of the Allies in France. The link-up of Operations Overlord and Dragoon in early September 1944 at Dijon was, in the words of Robert Citino, 'a strategic milestone for the Allies, a return to unity of effort and concentration of force'. It created a massive military force 'perhaps the most powerful in military history' stretching from the English Channel to the Swiss border. Rather than lasting a thousand years as Hitler promised, the Third Reich could not survive for much longer. In 1959, Truscott reflected on the importance of what Operation Dragoon had achieved:

> The Southern France operation did what it was expected to do for the main landing and in that sense it, of course, was worth doing. Had it not been done, what would have been the effect on those

German troops being free to withdraw and operate against the main attack up in the Rhine area later, is anyone's guess.

That a greater success had not been achieved in the south of France was not Truscott's fault. He had done everything he could to entrap the retreating German Nineteenth Army but lacked the military force necessary to achieve it. As Russell Weigley wrote of Truscott's performance here:

> No American commander drove harder than Truscott, and none clung most steadfastly to the principle that destroying the enemy army was the goal. Perhaps if he had led more than three infantry divisions that were constantly out running their supplies, he might have succeeded in the complete destruction of the enemy in his front which had constantly eluded the American generals.

In November 1944 Truscott was looking forward to a break from the war and a chance to catch up with his family. He later wrote that he could not recall the last time when he had felt so relaxed. He did not expect that his new appointment, primarily a training and administrative role as commander of an inactive 15th Army, would be as demanding or as stressful as what he had endured in Italy and France. Little did Truscott know as his aircraft touched down in the USA that the death of a senior British officer and Churchill's personal intervention would prevent him from assuming command of the 15th Army. Instead, he would soon be commanding a different army in a very active theatre of war.

7 Back to Italy

A new command

On 4 November 1944, Field Marshal John Dill, a former Chief of the Imperial General Staff, died of aplastic anaemia. Dill had been serving in Washington DC as the senior British representative on the Combined Chiefs of Staff Committee, an important diplomatic and representative post. Dill's sudden death set in train a major reshuffle of senior Allied officers which would eventually impact on Lucian Truscott Junior. In the military version of musical chairs that followed Dill's death, the Allied supreme commander in the Mediterranean, General Sir Henry Maitland 'Jumbo' Wilson, was sent to Washington DC to take Dill's place. Succeeding Wilson as the senior Allied officer in the Mediterranean theatre was Field Marshal Sir Harold Alexander, which led to Mark Clark's elevation to Alexander's previous role as the commander of the 15th Army Group in Italy. Clark's elevation left the command of his Fifth Army vacant. Lieutenant General Lucian Truscott Junior was the ideal replacement for Clark, although there were other contenders for the job. In fact, Clark had recommended one of his corps commanders, Major General Geoffrey Keyes, for the position.

On the last day of his leave Truscott called in to say farewell to General George Marshall prior to returning to SHAEF in France. Marshall stunned Truscott by asking him: 'How would you feel about going back to Italy?' Truscott gave the standard reply: 'Sir, I will go wherever you send me.' Marshall, annoyed not to get a direct response from Truscott, replied: 'That's not what I asked you. I asked you how you felt about it.' Truscott then admitted that he would prefer not to return to Italy.

The burdens of command. Truscott familiarizes himself with the situation in Italy prior to taking over command of Fifth Army. Note the ever-present packet of cigarettes on his desk.

He wanted to serve in the main theatre of war in Europe, which meant northern France, Belgium and Germany. He also wanted to be there at the final defeat of Germany, and he wanted to serve with General

Eisenhower again. Truscott also pointed out to Marshall that both corps commanders in the Fifth Army, Geoffrey Keyes and Willis Crittenberger, had been senior to Truscott all his career. 'Pay no attention to that,' Marshall snapped and then directed one of his staff officers to talk to Eisenhower about Truscott taking over the Fifth Army. What Truscott did not tell Marshall was that he did not relish fighting another winter campaign in Italy or serving under Mark Clark again.

When Truscott returned to France and met with Eisenhower, he learned that he would be proceeding to Italy forthwith. Eisenhower told Truscott that he was 'by all means, the person to go to Italy. . . . I hate to lose you, but I feel this is right.' Unknown to Truscott at the time was Churchill's intervention in the decision. The new Supreme Commander in the Mediterranean theatre, the newly promoted Field Marshal Harold Alexander, had made the initial request to Churchill to have Truscott brought back from France as the new Fifth Army commander. Churchill wanted Truscott too as he knew that Truscott got on well with the British and that Italy was familiar ground for him. Also, Churchill had not forgotten that it was Truscott who led the breakout from Anzio. Churchill raised the possibility of Truscott's return to Italy with President Roosevelt who agreed that it was a wise choice. Truscott's fate was sealed. When he met with Walter Bedell Smith at Versailles, Smith informed Truscott that there was not a chance of him remaining at SHAEF because 'The PM [Churchill] personally asked for you.' Against his wishes but determined to do the best job possible, Truscott was returning to Italy with the likelihood that this is where his war would end.

Fifth Army and Mark Clark

The new assignment as commander of Fifth Army meant that Mark Clark was again his superior officer as the Army Group commander. It was something that Truscott was anxious about given his previous

Return to Italy. Despite his reluctance, Truscott was sent back to Italy in December 1944 as the commander of Fifth Army. The II Corps commander, Major General Geoffrey Keyes, on the right, escorts Truscott around the frontline positions.

service with Clark. On meeting Clark again, Truscott found his concerns had not changed. He expressed his innermost feelings about Clark in a letter to his wife written on 11 December 1944. Truscott found Clark 'strange, most strange'. He elaborated that Clark had:

> plenty of ability but is one of the most self centered individuals I have known. He has always given me 100% support heretofore — and I had almost forgotten my inability to feel complete confidence in him — but it comes back in force. However, it will work out alright I'm sure.

Truscott arrived in Italy on 8 December 1944, bringing with him the loyal staff officers who had served with him in VI Corps and for his short time with the inactive 15th Army. The formal date for the changes in command was fixed by Field Marshal Alexander for 16 December. Truscott spent the days before assuming command visiting parts of the front line, meeting the senior officers in Fifth Army and assessing the condition of his soldiers. It was a different US Fifth Army that Truscott returned to. It consisted of three army corps, although one, the British XIII Corps, would be transferred to the British Eighth Army in January 1945. The other two corps were the US II Corps commanded by Geoffrey Keyes and the US IV Corps commanded by Crittenberger. Keyes' II Corps contained four US infantry divisions: the 34th, 88th, 91st and 85th Infantry Divisions. From January 1945 it also contained the Italian Combat Group 'Legnano'. Crittenberger's IV Corps was made up of two armoured divisions: the US 1st Armored Division and the 6th South African Armoured Division. It also had three infantry divisions: the US 92nd and 85th Infantry Divisions as well as the 1st Brazilian Infantry Division. From February 1945 the US 10th Mountain Division was also part of IV Corps. Most of the formations of Fifth Army now had considerable experience fighting the Germans in Italy.

The 'Buffalo' Division

The divisions with no combat experience were the 92nd Division and the 10th Mountain Division. In the segregated US Army, the 92nd Infantry 'Buffalo' Division consisted primarily of African American soldiers commanded by white officers. There was considerable doubt about the ability of these soldiers and the 92nd Infantry Division was to cause Truscott some concerns. The 92nd Infantry continued a long and proud tradition by adopting the Buffalo as its divisional symbol. More than 900,000 African Americans joined the US Army in the Second World War, all serving in segregated units and formations.

Truscott as Fifth Army commander talks with soldiers of the 92nd 'Buffalo' Division. While Truscott congratulated these soldiers on their recent performance, the 92nd Division, the only Army combat formation of African Americans in Europe, was to cause Truscott considerable problems during his return to Italy. Note the two sentries on the top of the building to the right.

Truscott decorates an officer he should have relieved of command. Major General Edward Almond, in sunglasses on the left, commanded the 92nd 'Buffalo' Division of African American soldiers. Almond has been described as 'an overbearing Virginian who would oppose integration of the armed forces until his dying day'. Almond once said: 'People think that being from the south we don't like Negroes. Not at all. But we understand his capabilities. And we don't want to sit at the table with them.' Almond was not the right person to command a division of African American soldiers.

Most were allocated to service rather than combat roles. The 92nd Infantry Division, originally formed for active duty in the First World War, was the only African American division to see combat in Europe. As such it was always closely scrutinized and its performance subjected to strong criticism. There is no getting around the fact, though, that the 92nd Division in Italy had a poor combat record until the very last months of the war.

When Lucian Truscott assumed command of the Fifth Army at the end of 1944, he was informed by General Devers that Mark Clark regarded the 92nd Division as 'wholly unreliable in combat'. Truscott, who had had a close relationship with two African American cavalry units before the war and regarded them highly, felt this was a harsh judgement. If the 92nd Division was assigned objectives within its capability, well-led and given adequate support, Truscott believed that this would lead to the Division achieving good results in combat. But when he tried to put this into practice in February 1945, in Operation Fourth Term along the rugged Serchio Valley, the 92nd Division's infantry regiments failed repeatedly to take and hold their objectives even against light opposition. Truscott concluded that the: 'Colored boys will not stay in position under any sort of fire, nor will their officers get close enough to the front to find out why they won't.'

General Marshall was in Italy and witnessed this attack by the 92nd Division. He later recalled:

> I was there (Serchio Valley) with Gen. Clark at the time of the fiasco of the 92nd Division, the Negro troops who simply refused to fight. Officers completely lost control over the men. The will to fight was completely lacking.

As a result of what he saw, Marshall concluded that the 92nd Division was 'untrustworthy under fire'.

Clark, Marshall and others had missed an essential point, although it was alluded to in the quote by Truscott above. They tended to forget

Napoleon's famous aphorism that: 'There are no bad regiments, only bad colonels.' As Truscott suggested, junior and senior officers of the 92nd Division were not forward with their men. Also, the commanding officer of the 92nd Division, Major General Edward M. Almond, had no love for nor did he respect the soldiers he commanded. Rick Atkinson has described Almond as 'an overbearing Virginian who would oppose integration of the armed forces until his dying day in 1975'. Almond attributed the failure of his division in combat to 'the undependability of the average Negro soldier to operate to his maximum capability, compared to his lassitude toward his performing a task assigned.... The general tendency of the Negro soldier is to avoid as much as possible.' This was a fatal attitude to have towards soldiers of a democracy. And not only does Almond's comment avoid his own command responsibility but it also lays the blame with those least able to change the situation and who look to their commanding officers for leadership and inspiration. As Truscott's biographer Wilson A. Heefner has written, the failure of Marshall, Clark and Truscott to hold Almond accountable for his division's failure or to remove him from command after repeated failures stands 'as a testimony to the racial mindset of [Almond's] superiors both in and out of the military'. Certainly, Truscott was remiss in not removing Almond from command and replacing him with a commander who believed in the abilities of the soldiers of 92nd Division.

Truscott did order an investigation into the performance of the 92nd Division since its arrival in Italy in August 1944. That report, conducted by Crittenberger, concluded that the division was unreliable and should not be used in offensive operations. A troubled Truscott forwarded the report to Mark Clark, stating that he concurred with its findings. He wrote: 'I think we must face the fact that we will never develop a combat unit capable of offensive power under our present system. What the answer is, I don't know.' Truscott's answer in Italy was to have the best African American soldiers, those decorated for gallantry, recently promoted or who had earned the combat infantry badge, placed into one regiment. The other two regiments of the 92nd Division were withdrawn to an inactive area and replaced by the 442nd Infantry (Nisei) Regiment and the 473rd Infantry Regiment which had been formed from converted anti-aircraft artillery battalions. The 442nd Infantry (Nisei) Regiment consisted of US soldiers of Japanese descent and the regiment distinguished itself in combat wherever it fought. In fact, this regiment became the most decorated of the Second World War with 21 Congressional Medal of Honor recipients. This solution worked well and the 92nd Infantry Division performed well in the spring offensive of 1945, earning more than 20,000 decorations and citations for bravery.

Truscott recognised that there was a serious problem with the performance of the 92nd Division, one that he did not know how or have time to fix. He later wrote that: 'Coloured soldiers were the product of heredity, environment, education, economic and social ills beyond their control — and beyond the sphere of military leaders.' The solution he came up with was the restructuring of the division so that it included only one African American regiment. He did not consider relieving Major General Almond, which would have been a solid start towards improving the morale and performance of the 92nd Division.

The Gothic Line

During Truscott's absence from Italy, the Fifth and Eighth Armies in Italy had advanced north from Rome to reach the Gothic Line. This was Germany's final defensive network south of Bologna set along the harsh mountain terrain of the northern part of the Apennine Mountains. The Gothic Line was not as formidable as the Gustav Line at Cassino, but it was still a strong defensive position consisting of the usual artillery observation posts, machine-gun nests, concrete bunkers and tank traps. Attacks made against the Gothic Line prior to Truscott's arrival had achieved some local success but not a decisive breakthrough.

This would not happen until the spring in April 1945 in the final Allied offensive in Italy.

When Truscott rejoined the Fifth Army as its commander in December 1944, he found it enduring a bitter winter at an altitude of over 2000 feet among the Apennine Mountains which were covered with ice and snow. He later recalled that: 'here, and ice, snow, mud, in fog we spent the winter'. His army was resting after a long, hard year and preparing for an offensive that would be delayed until the spring of 1945. His US Fifth Army faced the Gothic Line with the Ligurian Sea coast of Italy on its left flank. On its right flank, facing the Gothic Line from Bologna to the Adriatic Coast, was the British Eighth Army, which was stronger than Truscott's Fifth Army. It consisted of three corps and some sixteen divisions. Truscott's Fifth Army had been reduced since it had captured Rome in June 1944. At that time Fifth Army had numbered some 250,000 soldiers but over 100,000 of these had been sent to fight in France. At the end of 1944, Truscott's two corps of nine divisions numbered 160,000 soldiers, almost the same number he had commanded in VI Corps during the breakout from Anzio. Not one of Truscott's formations was trained in mountain warfare, although the 10th Mountain Division was expected to make up in part for the loss of the French Expeditionary Corps which had considerable experience in this type of fighting. The 10th Mountain Division consisted of skilled climbers, experienced skiers, including many students from Ivy League Colleges, and men used to operating in the mountains. Also included in its regiments were 500 Europeans, primarily from Norway. The 10th Mountain Division had spent months training at Camp Hale in Colorado. This camp had an elevation of 9224 feet with its training areas even higher. The 10th Mountain Division, some 14,000 specialized soldiers, reached Naples in mid-January 1945 much to Truscott's relief. After three years of solid training, the US Army's only mountain division was about to be put to the test. Facing the Allies on the Gothic Line were 27 German and six Italian Fascist divisions. These outnumbered the two Allied armies in Italy, but the Allies had complete control of the sea and air.

Truscott, as Fifth Army commander, presents his II Corps commander Major General Geoffrey Keyes with an award. The 15th Army Group commander, General Mark Clark, looks on. Truscott had worried about how his two corps commanders, both considerably senior to him, would take his promotion to command Fifth Army. In fact, Keyes had been one of Truscott's instructors on his officer training course in 1917. Truscott need not have worried. Both Keyes and Willis Crittenberger served Truscott to the best of their abilities.

Corps commanders

As commander of Fifth Army, Truscott had worried how the two corps commanders would receive his appointment as their superior officer. Both had been considerably senior to Truscott. In fact, Keyes had been one of Truscott's instructors at his Officer Training Camp in 1917 and Crittenberger had been one of his instructors at the Cavalry School. When Truscott raised the issue, General Marshall had abruptly dismissed his concerns. Marshall may have known something Truscott did not, as both Crittenberger and Keyes served Truscott loyally and to the best of their abilities. As Truscott later remarked, 'our relationships were always just as they should be', adding: 'We were friendly and there was never the slightest question of their complete loyalty.' The issue of Truscott's seniority never emerged during his time commanding Fifth Army in Italy.

A different affair

The day after Truscott assumed command of Fifth Army, a delegation from the US House of Congress arrived at his headquarters. These were members of the House Military Affairs Committee, and they were there to report on the progress of the Allies in Italy. After initial introductions were over, the Committee members were bundled into some 30 jeeps to visit locations of interest including the soldiers on the front line. While Truscott spent most of his time with Committee Chair, Representative John M. Costello of California, he could not help but notice the only female member of the Committee. This was the Representative from Connecticut, Clare Boothe Luce. The stunningly attractive Luce was already well-known in the United States as a writer, editor, journalist, novelist and playwright before she was elected to Congress as one of the early female representatives. Clare Boothe Luce was married to Henry Robinson Luce, the most powerful magazine tycoon in the United States who owned *Time*, *Life* and *Fortune* magazines among many others.

The marriage between Clare and Henry Luce was unusual. Henry had little interests outside of his publishing empire, was physically awkward and had no sense of humour. This was the complete opposite of his beautiful wife. Their marriage has been described as being 'sexually open' and Clare Boothe Luce certainly had an assortment of well-known lovers. Some of these included Joseph P. Kennedy, Randolph Churchill and the British author Roald Dahl. Clare had met Dahl in late 1942 when he was a fighter pilot in the Royal Air Force serving in Washington as

Clare Boothe Luce in 1932. Luce was an author and journalist who served in the US House of Representatives as a Republican from Connecticut from 1943 to 1947. She later served as the US Ambassador to Italy during the Eisenhower administration. Intelligent, articulate and very attractive, Clare Boothe Luce and Lucian Truscott had a brief affair while he was the Fifth Army commander.

an Assistant Air Attaché. Clare was obviously attracted to the dashing tall, athletic fighter pilot who was thirteen years younger than her. The British Embassy encouraged the liaison, hoping to convert Clare to a more pro-British stance. It is recorded in several sources that Dahl found the affair with Clare to be physically exhausting and reported to the British ambassador, Lord Halifax, in very blunt language, that he could not continue it. As Dahl's biographer, Donald Sturrock, recorded: 'Halifax apparently told him it was his patriotic duty to return to her bed.' Sturrock does point out that Dahl probably fabricated this story because it was 'hilariously funny', but there is no doubt that the affair took place. Despite Clare's sexual activities the Luces' marriage endured until Henry's death in 1967. One other name was to be added to the list of Clare Boothe Luce's list of lovers: Lucian Truscott Junior.

Why Lucian Truscott had this brief affair with Clare Boothe Luce is not hard to understand. The war was ending, Truscott was tired, lonely and he had been surrounded by death for so many years. He was also not in the best of health and must have felt the weight of his own mortality. Truscott's latest biographer, Harvey Ferguson, made the astute comment that, just as things looked so bleak for Truscott:

> Then, by accident or perhaps fate, a strong, vibrant woman came into his life. Just being in her presence sparked passions in him that had lain dormant for decades. What was especially pleasing was that she seemed to think so highly of him, making him feel younger than his years.
>
> He was a gone goose.

Clare had obviously been impressed with Truscott and the attraction was mutual. They began corresponding with one another and Clare arranged to visit Italy, this time alone, in March 1945. There is little doubt from Truscott's surviving correspondence that he and Clare had a brief but passionate affair that lasted less than two months.

It is probable that Sarah Truscott never knew of the affair.

The last German counterattack

On 18 December Truscott was summoned to Florence where Lieutenant General Mark Clark briefed him that the Germans were amassing troops and supplies for a counterattack in the Serchio Valley about 20 miles north of Lucca. The Serchio Valley, at the western end of the Gothic Line, had been a relatively quiet sector and was defended by fewer than two regiments of the 92nd 'Buffalo' Division. Facing them, Truscott knew, was a German and an Italian division, but Clark's intelligence revealed that these had been reinforced by a German mountain division, an SS panzer division and up to two more divisions. This was alarming information because if the Germans broke the 92nd Division's front, they could push on to threaten the vital port of Livorno (Leghorn) through which all of the Fifth Army's supplies arrived. Livorno was just 45 miles south of the Serchio Valley. Truscott immediately ordered two brigades of the 8th Indian Division and two regimental combat teams from the 85th Infantry Division to move to the valley and had the 1st Armoured Division move to Lucca as a fighting reserve. These formations were on the move when the Germans launched their attack.

Hoping to replicate the initial success of their counteroffensive in Belgium, France and Luxembourg, which was later designated the Battle of the Bulge, the Germans launched Operation *Wintergewitter* (Winter Storm) on 26 December 1944. It would later be called the Battle of Garfagnana and was to be the last German offensive in Italy. The 92nd Division was holding a six-mile section of the front from the coast to the village of Barga in the upper Serchio Valley. They were attacked by six battalions of infantry and mountain infantry supported by medium and light artillery. The Germans attacked from the town of Castelnuovo di Garfagnana and soon pushed the regiments of 92nd Division back with some units fleeing as much as five miles to the rear. In Truscott's words, some units of 92nd Division just 'melted away'. Three towns in the region were taken, including Barga. By 27 December the Germans had captured the area between Barga and the Serchio River, but this was the limit of their advance. Any further advances were held back by attacks from the Allied air force. The two Indian brigades moved up to four miles south of Barga on 26 December and by 29 December they had retaken the town. They also recaptured all of the villages and towns in the Serchio Valley that the Germans had captured. The German counterattack had not been pressed with the vigour of the attacks of the past and the small number of combat troops engaged indicated that their objectives were limited. It definitely did not match the German counterattack in France and Belgium, but Operation *Wintergewitter* achieved some success. It raised their morale for a short time. It had also forced the Allies to shift more troops to the west of the Gothic Line, delaying their planned attack on Bologna until the spring of 1945. The 92nd Division had performed poorly, but this was through a failure of leadership and not the courage of its soldiers, several of whom were decorated by Clark for bravery.

Stalemate on the Gothic Line

Holding the Gothic Line in the winter of 1944 was an ordeal for those who were there. The appalling weather and harsh terrain wore down soldiers' resilience and their morale suffered. It was not helped either by the fact that Italy was a secondary theatre of war with many of the soldiers there feeling forgotten or ignored. In an effort to keep morale high in Fifth Army, Truscott established rest centres in Florence, Rome, Capri and Sorrento. Only a small number of soldiers could be sent to these centres at any one time, but the men took heart knowing that their turn would come.

Truscott also took heart from a thoughtful Christmas message he received from General George Marshall. In a brief letter addressed to 'Dear Truscott' Marshall wrote:

This Christmas finds you again carrying the battle to the enemy in Italy. While I realize that you would have preferred to remain with our forces on the Western Front for the major effort, your vast experience with the fighting in Italy made it most advisable to entrust the command of Fifth Army to you. You have the complete confidence of the War Department.

Marshall also sent 'warm greetings' for Christmas and his 'best wishes for the coming year'. That the head of the US Army took time out from his busy schedule to show appreciation and gratitude of an officer's unselfish service added to Truscott's admiration and respect for General George Marshall.

Operation *Wintergewitter* in the Serchio Valley, the winter weather and harsh terrain had forced the Allies in Italy to postpone any large-scale offensive operations until the spring of 1945. Truscott, however, was not happy with just sitting on the defensive watching the offensive spirit of his Fifth Army drain away. He asked for permission to launch two limited attacks against enemy positions and was soon granted it. The divisions he chose to make these attacks were ones he wanted to test for their battle-worthiness. They were the 10th Mountain Division and the 92nd 'Buffalo' Division.

Testing the 10th

As mentioned above, the 10th Mountain Division was the US Army's only division shaped and trained for mountain warfare. As such it was not as strong as a regular infantry division and being 'new', no other theatre commander had wanted it added to their forces. To his credit, Mark Clark was keen to have any troops in Italy that would fight and readily accepted the 10th Mountain Division as part of Fifth Army. Truscott was equally pleased to have the division in his army from February 1945. He later wrote that it 'was one of the best combat divisions I knew during the war' and that its commander, Major General George P. Hays, was 'one of the ablest battle leaders I ever knew'.

On joining the Fifth Army, 10th Mountain Division was allocated a defensive position on the Gothic Line running from the Serchio Valley to Mount Belvedere. It was rugged, mountainous terrain that ran for a distance of 30 miles. Truscott's first task for the 10th Mountain Division was to capture the high ground on their right flank: Mount Belvedere. Belvedere was the first in a line of peaks stretching north-east along a three-mile ridgeline. The most prominent of these peaks was Mount Belvedere itself at 3900 feet. There were two other prominent peaks along the ridgeline at over 3000 feet. They were Mount Gorgolesco in the centre and Mount della Torraccia at the northern end of the ridge. These three mountain peaks overlooked Highway 64, one of the few roads to cut through the Apennines and one of the two major approaches to Bologna. Capture of this ridge would provide a solid platform to launch the offensive in spring and securing this ridgeline was essential for the Allies to make any further advance beyond the Gothic Line. Prior to Truscott's arrival, Fifth Army had made three attempts to take the Mount Belvedere ridge, but they were costly failures. This fourth attempt on Mount Belvedere was a test of the concept and effectiveness of the US Army's mountain division experiment.

Major General George Hays made a detailed reconnaissance of the area in which his division would be blooded, and his mountain knowledge helped him spot something that Truscott and other senior commanders had missed. Directly to the west of the Belvedere Ridge was another four-mile ridge ranging in height from 3200 feet to just over 4800. This was the Serrasicca-Campiano Ridge which the Fifth Army had named Riva Ridge. Hays deduced that while Riva Ridge was not as vital to the Gothic Line as Belvedere, it was being used by the Germans as an artillery observation post providing essential artillery protection for that key feature. While the German garrison on Riva Ridge was only about 50 men, their well-sited machine-gun positions were enough to deter any attack on the ridge. The capture of Riva Ridge needed to be part of the plan for the assault on Belvedere Ridge.

For weeks before the attack, patrols from 10th Mountain had scouted up to both ridges, marking routes, climbing the sheer rock faces and installing ropes, pitons and anchors. For both ridges Hays had selected the most difficult ascents, knowing that these would be lightly guarded. To ensure that surprise was obtained, the climb to the assault had to be made at night with weapons unloaded. For weeks the Fifth Army had shone bright lights onto both ridges. The lights were necessary to blind the German defenders and to provide some light for the 10th Mountain Division climbers. At first the Germans were suspicious and increased their vigilance. However, after weeks of bright lights at night the Germans became used to it and relaxed their guard.

On the night of 18 February 1945, the 1st Battalion of the 86th Mountain Regiment commenced its climb on Riva Ridge. Some 700 soldiers of the battalion had to climb up 2000 feet of vertical cliff to reach its top. A fog set in during the night, aiding their concealment but making the climb more difficult. Reaching the summit, the mountain soldiers found the observation posts unmanned and occupied them. They then attacked the sleeping Germans with grenades and rifle fire. The German garrison was easily overcome.

For the rest of the night soldiers of the 10th Mountain Division, using pack animals, moved ammunition, rations and heavy weapons on to Riva Ridge to prepare for the inevitable German counterattack. The weapons included four .50-calibre machine guns and a pack howitzer. The German counterattack was launched at 10 a.m. on 19 February and it lasted until nightfall. That evening, with the attack on Belvedere Ridge about to commence, the Fifth Army used Riva Ridge to direct its own artillery in support of their attack. The Germans tried to recapture Riva Ridge for the next five days without success.

On the night of 19 February six battalions from the three regiments of 10 Mountain Division, some 13,000 men, made their assault on the Mount Belvedere Ridge. After the attack on Riva Ridge the Germans were now alert and ready for another assault. The fighting was brutal and while Belvedere's summit was taken on the morning of 20 February, it took the 10th Mountain Division six days before all three peaks of the ridge were in their hands. The last attack was made on Mount della Torraccia on the evening of 24 February with much of the fighting hand to hand. The German defenders surrendered the next morning.

The 10th Mountain Division proved its worth in this attack on the two mountain ridges. Truscott was pleased with their performance, noting that although this was their first fight, the mountain soldiers had 'performed like veterans'. He had expected the 10th Mountain Division would need two weeks to capture both ridges. The division did it in five days. It had not been a cheap victory, though. Casualties in the division numbered over 900 with 192 being killed in action and a further 730 wounded. One regiment, the 85th Mountain Infantry, which had been involved in the initial assault on Mount Belvedere and then pushed on to take Mount della Torraccia and two hills in between, had suffered more than half these casualties with 470 dead and wounded. Total German losses were unknown but over 400 had been taken prisoner.

Restructuring the 92nd

The limited attack by the 92nd Division was made on 8 February in the Serchio Valley and Cinquale Canal area. Its three-day attack achieved nothing but heavy casualties numbering more than 500. Another 200 men had deserted before the attack and needed to be rounded up. Truscott reported the failure to Mark Clark, stating: 'The division has been satisfactory in every respect except the one element which justified its existence — the combat infantry.' It was after this failed attack that Truscott reorganized 92nd Division. It would have one regiment of African American soldiers, one of soldiers of Japanese origin and one regiment of artillery soldiers now converted to infantry.

Planning the last offensive

As the welcome spring thaw in the Apennines arrived, both Allied armies in Italy started planning for offensive operations. These would be hampered somewhat by the loss of several formations to other theatres of war. In February 1945 Mark Clark had been alarmed to learn that he was to lose five Eighth Army Divisions including the much vaunted and experienced Canadian Corps. They were to be sent to northern Europe to bolster Montgomery's massive 21 Army Group. Facing Mark Clark's 15th Army Group in Italy were 24 German and five Fascist Italian Divisions with seventeen of these holding the Gothic Line. Of Clark's two armies in Italy, the British Eighth Army now numbered eleven divisions and some Allied Italian units. Truscott's Fifth Army contained nine divisions, seven of which were American. The other two formations were a Brazilian and a South African division. There was also an Allied Italian unit and some other small units as part of his army. The bulk of the fighting, though, would fall on the infantry and armoured divisions. While the two armies of 15th Army Group in Italy were outnumbered by the German defenders, the Allies had absolute air supremacy and they ruled the waves around Italy's shores.

Mark Clark's plan for the final Allied offensive in Italy, codenamed Operation Grapeshot, had three phases. It would begin with a diversionary attack by the restructured 92nd Division north along the Ligurian Sea coast to capture Massa and the Italian naval base at La Spezia. Four days later the main attack was to commence with the Eighth Army breaking through the Gothic Line and driving north and north-east. Three days later Fifth Army was to launch its main attack, proceeding north and north-west. Both armies had full air support during the offensive.

If it succeeded, Clark's two armies would emerge from the Apennines which had penned them in and advance into the Po Valley. The city of Bologna would be taken and the remaining northern third of Italy liberated from the Germans and Italian fascists. The destruction of all Axis forces in southern Europe would be complete. For this final offensive Truscott had over 200,000 soldiers, 2000 artillery pieces and thousands of vehicles positioned along a 120-mile front which ran from the Ligurian coast to the city of Bologna. Tasked with starting the offensive, the 92nd Division had to fight a separate battle very much on its own. Truscott's IV Corps, consisting of 1st Armored, 10th Mountain and the Brazilian divisions, were the left flank of the offensive. On their right, II Corps, with the 34th, 88th, 91st Infantry divisions, the 6th South African Armoured Division and an Italian Group were in the centre of the offensive and had the task of capturing Bologna and linking up with Eighth Army. Truscott held one division, the 85th, in reserve.

The final offensive in Italy

The last Allied offensive in Italy commenced with the attack by the 92nd Division on 5 April 1945. Although only meant as a diversion, 92nd Division's attack achieved total success. The town of Massa was captured two days after the attack commenced, then nearby Carrara on 11 April. But the 92nd Division did not stop there. It pushed on, battling bad weather, steep mountains and determined German resistance to capture the important port city of La Spezia on 25 April. The division kept advancing further north along the Italian Riviera, reaching Genoa on 27 April to find that the city had been liberated by Italian partisans. Some 4000 Axis soldiers were taken prisoner at Genoa. Truscott was pleased with the progress of 92nd Division, later writing that: 'As a diversionary measure it was wholly successful.'

As the 92nd Division was fighting its way north to Carrara, further east the Eighth Army commenced its attack on the Gothic Line as scheduled and made steady progress. Bad weather, keeping all aircraft grounded, delayed Truscott's Fifth Army assault until 14 April. In the battle through the Apennines the 10th Mountain Division led the way. On

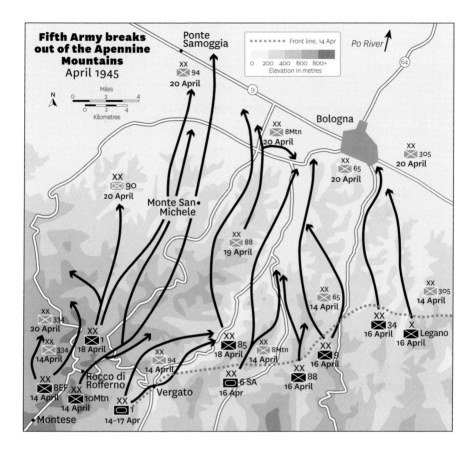

14 April Truscott watched as this division captured the Rocca di Roffeno massif in the face of stiff enemy resistance. As it did so, Truscott ordered the 1st Armored Division on the right flank of 10 Mountain to start its tanks moving north. By the end of the day 1st Armored Division had captured the town of Vergato after fierce house-to-house fighting. On the 10 Mountain Division's left flank the Brazilian Division captured the town of Montese. Truscott was effusive in his praise of the mountain soldiers, later writing: 'The 10th Mountain Division would carry the brunt of the attack to the Po Valley — and beyond.'

Fighting continued over the next five days as Truscott's IV Corps pushed the German defenders from one ridgeline to another. Casualties were heavy on both sides, but the German defenders, whose morale had not been good at the start of the offensive, were close to breaking point. By 19 April the Fifth Army had advanced six miles into the Apennines.

The day after IV Corps' advance, II Corps, on their right flank, began its attack supported by over 1000 bombers. With this massive air support, the South African 6th Armoured Division and three infantry divisions pushed steadily north astride Highways 64 and 65 towards Bologna. They reached the outskirts of the city on 20 April.

Meanwhile, on the left of IV Corps, as 10 Mountain Division continued its push north, Truscott had moved 1st Armored Division to the left of the mountain soldiers where the ground was more suited to tanks. His reserve division, the 85th Infantry Division, took over 1st Armored's original sector. Free of the mountains in the Samoggia River valley, 1st Armored Division's Sherman tanks drove along the valley destroying all enemy opposition, including the few tanks the Germans could muster. Their Gothic Line defences were disintegrating. As the American tanks were creating havoc in the river valley, 10th Mountain Division advanced another six miles and seized the Monte San Michele peak, capturing almost 3000 Germans. The next day the 10th Mountain Division pushed on to reach Ponte Samoggia across Highway 9 and to now be clear of the Apennines.

By 20 April 1945, Operation Grapeshot had fundamentally changed the situation in northern Italy. Both Fifth and Eighth Armies were beyond the hostile terrain that had held up their progress for so many months and the Germans were on the run. Truscott's Fifth Army was finally beyond the Apennines and into flat open country that favoured rapid manoeuvres.

Truscott's II Corps was now ordered to encircle Bologna and link up with the Eighth Army at Bondeno, 20 miles north of the city. Truscott's IV Corps was to continue its drive to the north past Modena to get across the Po River before the Germans reached it. The 34th Division in II Corps captured Bologna on 21 April. It turned the city over to the Italian Legnano Brigade and pushed on north-west to Modena. Meanwhile the South African armoured division and the US

88th Division, which had enveloped Bologna to the west, pushed on to reach Bondeno. Further west, IV Corps advanced into the Po Valley along a broad front. By 23 April the divisions in IV Corps had reached the southern bank of the Po River and the Allies held the riverbank from west of San Benedetto to the Adriatic Sea. Both Fifth and Eighth armies had linked up at Bondeno, isolating all the German formations trapped east of Bologna. In Truscott's IV Corps, the Brazilian soldiers and the US 34th Division advanced west along Highway 9 towards Piacenza, isolating those German formations in the western portion of the Apennines. In all, some 100,000 German soldiers were captured south of the Po River by this rapid Allied advance.

The Po River, being the longest river in Italy and more than 100 yards wide in places, was a formidable obstacle. In April 1945, all bridges across it had been destroyed. However, on 24 April 10th Mountain Division, using 50 M2 assault boats, ferried two regiments across during the day and the remaining regiment that night. The 88th Division also established two bridgeheads across the river on 24 April and both 91st Infantry and the South African 6th Armoured Division were across the river the following day. By 26 April, Army engineers had built pontoon and foot bridges across the Po River with tanks, artillery, heavy equipment and reinforcements being able to cross.

There was to be no pause after crossing the Po River. The Allies worried about the Germans establishing an Alpine Redoubt on the Adige Line, so Mark Clark directed his forces further north to prevent this. The 10th Mountain, 85th and 88th divisions were directed to advance on Verona, which was captured on the morning of 26 April by 88th Infantry Division after a fierce night battle. The 88th Division pushed on north towards Vicenza which it also captured on 28 April. Meanwhile the 91st Division had crossed the Adige River at Legnano. Linking with the Eighth Army as it advanced towards Padua and Venice and protecting Fifth Army's right flank was the South African 6th Armoured Division. Further west the 10th Mountain Division advanced along the eastern shores of Lake Garda, while the 1st Armored Division secured the area between the lake and Como. The 34th Infantry

Division had also been active, capturing the towns of Parma, Fidenza and Piacenza. Truscott's Fifth Army had now captured more than 80,000 enemy soldiers.

German surrender

The German defenders had no answers to this rapid Allied advance in northern Italy. Their defeat was inevitable. On 18 April German delegates arrived at Mark Clark's 15th Army Group Headquarters at Caserta to negotiate a surrender. They signed the surrender documents the next day with the official date of their capitulation to be 2 May 1945. The official ceremony took place at Clark's Headquarters on the afternoon of 3 May. The surrendering German generals had first passed through Truscott's headquarters, but he had refused to meet with them, sending them on to Clark's headquarters without delay. The next day Truscott was ordered to attend a staged photograph session at Florence along with the commander of the British Eighth Army, Lieutenant General Richard McCreery. The Germans were represented by General Fridolin von Senger und Etterlin who had defended the Cassino position so well the previous year. The photograph taken shows von Senger saluting Clark with Truscott standing at attention wearing his well-worn cavalry trousers and leather jacket. He looked tired and unimpressed. Truscott later noted: 'The meeting struck me as pointless for the only purpose was a photographic record in the Hollywood tradition.'

Truscott was pleased and proud of his Fifth Army's performance in its last campaign. He told an interviewer in 1959 that: 'Our final campaign was one of the really outstanding campaigns of the war, I think.' Certainly, Mark Clark was pleased with both Fifth Army and Truscott's performance as its commander. Writing to Truscott on 25 May that his army 'excelled in the spring offensive which ended in the German capitulation on May 2nd', Clark was effusive in his praise:

No Army ever deserved a great victory more than the Fifth, and I am happy and proud that it was part of the 15th Army Group which was the first Allied unit to which the Germans surrendered. I therefore desire to commend the troops of your command for their outstanding achievements and you personally for your superior leadership.

In an assessment of Truscott's performance written on 24 May, Clark described him as having 'a thorough knowledge and quick grasp of military problems, particularly in combat and makes sound and rapid decisions'. Further to this Truscott had 'a quiet but forceful personality, coupled with obvious personal courage and determination, which inspires confidence and loyalty of all personnel associated with him'. Clark had obviously changed his opinion about his 'whining' subordinate. He now rated Truscott as Number 1 from 71 generals.

The end in Italy

The Italian campaign had lasted more than twenty months and saw some of the hardest fighting of the war. Some 500,000 Allied soldiers fought in Italy as they slowly advanced more than 1100 miles up the country. Of 312,000 Allied casualties, 189,000 were suffered in the US Fifth Army. Some 20,000 American, 6600 British Commonwealth, 5000 French and 48,000 German soldiers died during the fighting in Italy as did countless Italian civilians.

Truscott felt the losses keenly as evidenced by a Memorial Day speech he made at the Nettuno military cemetery at the end of May 1945. This military burial ground had been opened just two days after the Anzio landing. Returning to the Anzio beachhead and to its burial site was a poignant moment for Truscott. What was once a muddy field has now been transformed into the Sicily-Rome American Cemetery, a 77-acre sanctuary where 7845 American soldiers, sailors,

Truscott mixes easily with the soldiers who were chosen as the most outstanding members of their respective divisions during a banquet at Fifth Army headquarters in Gardon, Italy, in August 1945. The honoured men, all non-commissioned officers, are, left to right: William R. Abrams, 88th Division; John L. Daniel, 85th Division; Truscott; William H. Eastland, 34th Division; Rosario V. Larma, 91st Division; and Troy J. Perks, 92nd Division.

nurses and airmen have their final resting place. A further 3095 names of the missing are now engraved on the white marble walls of the chapel there.

On 30 May 1945, with many of the graves freshly dug, Truscott climbed to the speaker's platform and surveyed the crowd in front of him. It included senior officers, representatives from Congress, local VIPs, reporters and ordinary soldiers paying their respects. He then did something remarkable, acknowledging those seated in front of him before turning his back and speaking directly to the dead Americans, casualties of the bloodiest event in human history. Truscott spoke without notes in his distinct gravelly voice. According to Sergeant Bill Mauldin, the famous US war cartoonist who witnessed it, the speech:

> came from a hard-boiled old man who was incapable of planned dramatics. The general's remarks were brief and extemporaneous. He apologized to the dead men for their presence here. He said everybody tells leaders it is not their fault that men get killed in war, but that every leader knows in his heart that this is not altogether true. He said he hoped anybody here through any mistake of his would forgive him, but he realized that was asking a hell of a lot under the circumstances. . . . he would not speak of the glorious dead because he didn't see much glory in getting killed in your late teens or early twenties. He promised that if in the future he ran into anybody, especially old men, who thought death in battle was glorious, he would straighten them out. He said he thought that it was the least he could do.

Mauldin recalled that what Truscott did on this Memorial Day was 'the most moving gesture I ever saw'. Truscott's military aide wrote in his journal of this event: 'A beautiful Memorial Day ceremony and the Army Commander outdoes himself in his speech.' It took a special individual to do such a thing. It was perhaps Lucian Truscott's second finest hour.

 After the war

Taking over from George Patton

Six days after the signing of the German surrender in Italy, the rest of the German military surrendered across Europe. After this, the various formations that had made up Truscott's Fifth Army were assigned other roles outside of his command. The headquarters of Fifth Army was made non-operational in September 1945 and what remained of it was deactivated back in the USA on 22 October. Truscott was not there to witness the final demise of his army in what must have been sad occasions for him. In June 1945 he had returned to the USA to take part in various victory celebrations and to seek a new assignment from General George Marshall. Truscott had made it known to him that he was available to serve in the Pacific war.

On 29 July Truscott had his new assignment from Marshall. He was to go to China where he would command a group of armies fighting against the Japanese. Truscott and his staff were on their way to China on 8 August 1945, the day Japan surrendered. From China, Truscott and his staff travelled back to Italy to pack up their headquarters and personal belongings before returning home for the final time. The journey would be from Paris, which left Truscott just enough time to say goodbye to old friends including Patton, Eisenhower, Geoffrey Keyes and 'Beetle' Smith. Little did Truscott know that a visit to Eisenhower in Frankfurt would delay his return to the USA by more than a year.

Eisenhower had a new assignment for Truscott. He needed to replace George Patton as commander of the Third Army and military governor of the Eastern Military District (Bavaria). Patton had finally burned his last bridge by a public statement that members of the German Nazi Party were no different to those who joined the Democratic or Republican parties in the United States. The comment caused outrage in the United States and Eisenhower realized that he needed to remove Patton before he caused more damage to the reputation of the US Army. Eisenhower summoned Patton to his headquarters in Frankfurt to inform him he was being relieved of command of Third Army. When Eisenhower asked Patton who should take over from him, Patton replied instantly, 'Lucian Truscott.' Little wonder that when Truscott called into Eisenhower's headquarters to say farewell he was greeted with: 'Lucian, you are just the man I need.'

Taking over from Patton, a man he respected and liked, was a difficult decision for Truscott. Not aware that Patton had recommended him, Truscott wrote of it:

> General Eisenhower knew, of course, that Patton was my close personal friend. I explained that I had no desire to supersede him, but that I wished to be of service. If General Patton had to depart, I thought he would probably prefer being replaced by me than by someone who might be less sympathetic.

Certainly, many elements of the US media had no sympathy for Patton's demise. Under a headline of: PATTON AND HIS PISTOLS SENT PACKING, one reporter wrote:

> Gen. George S. Patton, the brilliant military commander who can't keep his mouth shut or his hands to himself and who can't tell the difference between Nazis and Republicans and Democrats, will finish his career behind a desk in command of a phantom army.

Patton insisted on a formal exchange of command. Both men must have been aware of the irony of the situation. Truscott was replacing the same army commander who had threatened to relieve him in Sicily for a perceived excess of caution and lack of aggression. Patton, in his turn, was now to command the inactive 15th Army headquarters tasked with collecting material for the official history and compiling lessons learned from the war. The 15th Army had been Truscott's first (inactive) army command prior to his return to Italy. Now it was Patton who would serve the US Army from behind a desk. Patton would not do this new job for long. On 9 December he was seriously injured in a car accident that left him paralyzed. He died in his sleep on 21 December 1945. Truscott attended the funeral of his old friend on Christmas Eve and commanded the military escort to the American cemetery at Hamm in Luxembourg.

The military governor

Replacing Patton as military governor of the Eastern Military District was a huge task. The district included all of the state of Bavaria and a part of Czechoslovakia. In this district Truscott's task was to disarm the German military, reconstruct the civilian institutions and punish those responsible for war crimes. It involved rebuilding civilian administrations without any members of the Nazi Party and resettling displaced persons of which there were over a million in the American sector alone. All of this had to be done as Truscott's Third Army shrunk in size. Those who remained on occupation duty had little liking for the task and there were many incidents of unruly behaviour. Truscott later reflected that: 'What had been a magnificent fighting force became little more than a rabble — an undisciplined mob.'

One of Truscott's many duties was the administration of the war crimes prosecutions in Nuremberg and Dachau. He was determined that those identified as Nazi war criminals should receive a fair trial. For the Nuremberg trials Truscott placed Brigadier General Leroy H. Watson in charge and made sure he had all the resources he needed to get the job done. While Truscott knew that the war crimes trials were necessary, he found the process distasteful and could not escape the feeling that he was taking part in a victor's justice. He wrote of it in *Command Missions*:

> On my visits to Nuremberg, I had many discussions with officials connected with the trials. I was present on the opening day and heard the reading of the indictments. At various times, I listened to the presentation of the testimony. I saw the accused in their cells, exercise in the prison yard, in conference with their counsel, and before the tribunal in the courtroom. It was one of imposing dignity, but it was not one to fill me with the pride I have felt in American courtrooms. I believed that these major Nazis were guilty of waging aggressive war and other crimes against humanity for which they should be brought to the bar of justice. But when I looked down upon the courtroom scene I was never able to escape the impression that I was witnessing a conquerors' triumph, for it was only the totality of the conquerors' victory that made this impressive spectacle possible.

Most of those on trial at Nuremberg were Nazi leaders and nearly all were found guilty. They received sentences ranging from years in prison to death by hanging.

The war trials at Dachau were for what Truscott called 'the lesser fry' of the Nazi regime. It was specifically for Nazis who had committed war crimes against American personnel or committed them in the American sector of occupation. Truscott was more comfortable with these trials as they were easier to conduct away from the publicity of Nuremberg and they were run by the United States military. Instead of a jury, the American military tribunal set up a panel of seven men, one of whom was an expert in international military law. Those on trial were concentration camp officers, guards and medical personnel, members of SS units, those who had tortured or murdered downed

Allied pilots and those responsible for the massacre of US soldiers at Malmedy in Belgium. Of the first 40 Nazi officials tried, 36 were sentenced to death and Truscott signed their death warrant. It was the only one he signed. Of the 36 sentenced to death, 28 were hanged on 28 and 29 May 1946, including Dachau's former commandant and camp doctor.

Truscott's time as military governor was not just devoted to crime and punishment, although this took up a lot of his time. A more pleasant task for Truscott, a former educator, was the opening of a university for the displaced persons in the American sector. Sponsored by the United Nations and supervised by the Third Army, the university's first student cohort consisted of 1300 students, about half the number who had applied. Based in Munich, Truscott signed the university's charter on 16 February 1946.

Forced retirement

In April of 1946 Truscott learned that his wife Sarah was seriously ill and had been admitted to Walter Reed General Hospital in Washington DC. Truscott flew back to the US to be with Sarah and stayed with her until she recovered. However, on the return flight to Germany Truscott himself fell ill with a serious chest infection. At a hospital in Heidelberg doctors treated the chest infection with penicillin, but an electrocardiograph machine picked up a heart condition. Truscott was ordered to undertake six weeks' bed rest immediately. His old friend Geoffrey Keyes was coming from Italy to take over the command of Third Army. Truscott wrote with some bitterness: 'That ended my second Army command and my last wartime mission.' In hospital, Truscott was also advised to give up smoking and to cut down on his alcohol consumption. While he could not smoke in the hospital, Truscott was able to keep drinking by arranging to have two highballs served before lunch and two more before dinner.

The strain of the war years is clearly evident in this photograph of Lieutenant General Lucian Truscott taken in April 1946 when he was serving in Germany as commander of the Third Army and military governor of Bavaria. Just weeks later, Truscott fell seriously ill with a chest infection and a heart condition. It was the end of his active military career.

After the six weeks in the Heidelberg hospital, Truscott returned to the United States in July to be placed under six months of observation by doctors at Walter Reed Hospital. Following this period of observation, Truscott spent six months serving with the War Department Personnel Boards while awaiting the results of a final medical examination. The results were not good. It confirmed a serious heart condition and concluded: 'This office is of the opinion that General Truscott may be incapacitated for active service.' He was retired from active duty on 30 September 1947, aged 52. He had served in the US Army for 30 years and six weeks.

For almost four years Lucian and Sarah Truscott lived on a 70-acre country estate near Bluemont, in Loudoun County, Virginia. It was excellent horse country, and the estate had several sheds, one of which Lucian used for his woodwork projects. Lucian Truscott found it peaceful but missed the action. A four-month recall to active duty by General Devers working on US Army doctrine offered a brief respite, but it was only temporary. He started gathering material for two books, but this was interrupted when another old friend offered him a job with a promise of action.

Working for the CIA

In early 1950, President Harry S. Truman appointed retired general Walter Bedell (Beetle) Smith as head of the Central Intelligence Agency (CIA). It was hardly comforting to Smith, who had served a term as ambassador to the Soviet Union, that he would be the CIA's fourth director in four years. It was Smith who offered Truscott a new job. In February 1951, Truscott was hired by the CIA, first as a consultant, then as the CIA's senior representative or Station Chief in Frankfurt, Germany. When 'Beetle' Smith retired from the CIA in 1953 due to ill health, the new director, Allen Dulles, kept Truscott on.

Truscott, with Sarah in tow, returned to Germany in 1951 working for the CIA, although his official title was 'Special Consultant to the United States High Commissioner'. Frankfurt was a very active CIA post being the front line of the new Cold War. While in his role as Station Chief in Germany, Truscott shut down an illegal CIA prison, oversaw the digging of a tunnel from West Berlin to the east in order to tap East German and Soviet telephone lines, foiled the planned CIA assassination of Chou En-lai of Communist China and wrote a report for Eisenhower on the CIA's involvement in the failed Hungarian uprising of 1956. As a result of Truscott's report, Eisenhower ended the CIA's activities in Czechoslovakia where it was also promoting unrest.

In 1956 the CIA awarded Truscott its Distinguished Intelligence Medal and in the following year it promoted him to deputy director. His new role required him to adjudicate on disputes between the CIA and military intelligence. Truscott was deputy director of the CIA for two years. In May 1959, ill health forced him to resign.

Two books

In 1954 Lucian Truscott Junior published *Command Missions: A Personal Story* which dealt with his assignments during the Second World War. It was well received, and it is still highly regarded. The Modern War Institute, for example, when reviewing the book recently stated of *Command Missions*: 'This volume is by far the finest memoir written by an American field commander from World War II.' It was typical of Truscott that he ended *Command Missions* not by reflecting on his own career but instead paying a touching tribute to the qualities of the American soldier who when 'properly equipped, trained and led, he has no superior among all the armies of the world'.

Truscott's other publication, *The Twilight of the U.S. Cavalry: Life in the Old Army, 1917–1942*, was incomplete when Truscott died. The book was a touching tribute to the horse soldier between the wars and records with considerable sadness the demise of the US Cavalry during that period. Truscott's son, Lucian III, completed the book and wrote the 'Editor's Preface' which reveals much about the character of his father. It was published in 1989 by the University Press of Kansas. Lucian Truscott Junior barely features in *The Twilight of the U.S. Cavalry* but records events as an eyewitness to this piece of American history. One of the few times Truscott was mentioned by name was at the end of the 'Author's Preface' where he signed off as 'L.K. Truscott, Jr, Cavalryman'.

Seven months after the publication of *Command Missions*, President Eisenhower signed off on an Act of Congress that promoted some generals to the retired list. This Act promoted Truscott from Lieutenant General to a full four-star general. In some ways it was a hollow achievement as part of the Act stated: 'No additional retired pay accrues as a result of this advantage.' Truscott did receive a new 'General Officer' flag with the four stars and an equivalent automobile pennant.

Family relationships

Lucian III recalled that his father mellowed as he aged. He described one incident that had the potential to be very unpleasant but gave him some insight into his father. It happened shortly after the war when Lucian III and his wife had just returned from a tour of duty in Japan. Sarah Truscott had organized a social occasion involving many family members and friends. As coffee was being served after dinner, Lucian Truscott took out a cigarette from a silver box near at hand. Lucian III immediately took out his lighter, snapped it alight and held it up for his father. Lucian Truscott looked at him and said: 'Son, something you've got to learn is that one of the first things I always taught my damned aides was that I can light my own cigarette.' Without thinking, Lucian III immediately replied: 'Dad, something you've got to learn is that

I'm not one of your damned aides.' For a moment there was stunned silence at the dinner table. Lucian Truscott looked at his son as the lighter burned 'for what seemed an eternity' then he grabbed Lucian III's wrist, gently pulled it towards him, and lit the cigarette. Lucian III reflected that both of them learned something that night: Truscott, that his son was a grown man and not one of his subordinates; Lucian III, that the elderly man beside him 'was someone besides that domineering father image of my youth'.

Lucian III also wrote that his father 'was certainly a pushover for his grandchildren in the few years immediately before he died'. He spent considerable time with his grandchildren teaching them woodwork and games. A 'pushover' did not mean he let his standards slip. One of his grandchildren, Mary Truscott, recalled receiving a gentle rebuke at the dinner table delivered in verse in her grandfather's gravel-like voice:

Mary Truscott,
strong and able
this is not a horse's stable
get your elbows off the table.

Not just Mary, but every other grandchild present, received the message that some standards needed to be maintained no matter how much their grandfather doted on them.

Last post

Leaving the CIA in 1959 was Truscott's last retirement. After it, Lucian and Sarah retired to a 'cottage on the Potomac' in Waynewood south of Alexandria, Virginia. Truscott's health in his final years was not good. His war service in the European winters caught up with him and was not helped by his heavy smoking and drinking. Truscott's ailments included stomach ulcers, oedema (a condition in which fluid accumulates in the body), emphysema, gout, and heart problems. General Lucian Truscott Junior died in Walter Reed Hospital on 12 September 1965. The cause of death was listed as 'pulmonary emphysema, chronic lung disease, pulmonary insufficiency, and myocardial insufficiency'. He was 70 years old.

9 Assessment

What people said

There is little doubt that Lucian Truscott Junior was one of the most important Allied senior commanders of the Second World War. There is also broad agreement about how well Truscott performed in his various command appointments during the war, especially as commander of an infantry division and higher. Featured below are some judgements about Truscott's command skills. They are made by both his contemporaries and by military historians.

> 'Truscott was one of the really tough generals. He could have eaten a ham like Patton for breakfast any morning and picked his teeth with the man's pearl-handled pistols. Truscott spent half his time at the front, the real front, with nobody in attendance but a nervous jeep driver and a worried aide.'
> — Bill Mauldin, Pulitzer Prize-winning editorial cartoonist and veteran of US 45th Infantry Division

> 'He is one of the best Army commanders, second only to Patton. Experienced, balanced fighter, energetic, inspires confidence. Since the landing in North Africa in the fall of 1942, this tough, hard-driving general has achieved outstanding success as a battlefield commander in a succession of regimental combat team, division, corps, and field army.'
> — General Dwight D. Eisenhower

'Truscott — shy in a crowd — a handsome man — a fine combat commander who inspired the confidence of his troops and of higher commanders — dynamite.'
— Field Marshal Sir Harold Alexander

'I think Truscott is the finest army combat commander on the battlefield today.'
— General Jacob L. Devers

'Lucas was replaced by Truscott, who was a damn fine soldier.'
— General Hamilton Howze

'He was a hard fighter. We had our little disagreements, but I had great confidence in his tactical ability and aggressiveness.'
— General Mark Clark

'In those days, I thought I had quite a talent for fighting.'
— Lieutenant General Lucian Truscott Junior

'He had toughness, courage, tactical ability and professional competence. He also had an intangible only the best possessed: great leadership under fire — the genius for doing what must be done in the heat and chaos of battle that separates the adequate from the exceptional.'
— Carlo D'Este

'Many considered him the finest combat commander in the U.S. Army.'
— Rick Atkinson

'He was the very model of a can-do US Army officer.'
— Robert M. Citino

Unique experience

Truscott's experience of the Second World War was unique. He was the only senior US commander to command at all senior levels during the Second World War. This included a regimental combat team, an infantry division, a corps in Italy and France, and a field army.

Starting with the regimental combat team in Operation Torch, the fight to capture Port Lyautey and the Kasbah had been chaotic, and events had not unfolded to plan. Truscott's command had been satisfactory but hardly more. His opening line of *Command Missions*, his account of the war, was the simple statement that: 'Luck plays a part in the life of every man.' It certainly played a part in Truscott's rise during the Second World War. Truscott was lucky that Eisenhower was both his friend and patron, one who was concerned that Truscott be given every opportunity to succeed at higher levels of command. It was also lucky for Truscott that, after serving briefly as Eisenhower's deputy in North Africa, he was assigned to command the 3rd Infantry Division.

It was as commander of this division that Truscott came into his own. He was a hard-driving stern commander who cared deeply about the dogface soldiers he led. Under Truscott the 3rd Infantry Division was soon recognized as one of the best in the US Army. It distinguished itself in the fighting in Sicily and on the Italian mainland.

Truscott's next command, VI Corps at Anzio, was his finest hour. He transformed the Anzio beachhead from one that was constantly looking over its shoulder to one that knew it had the measure of the Germans and looked forward to the eventual breakout. In the process Truscott restored the faith of his British formations in their American Corps commander. Preparing for the breakout from Anzio saw him in command of a very large corps, some 150,000 soldiers. It was Truscott who planned and controlled the breakout from the Anzio beachhead, and when it came, it was spectacular with the potential to entrap an entire German Army. That it did not do so was beyond Truscott's control, when he was ordered to abandon the agreed plan and focus on the capture of Rome instead.

Commanding VI Corps in France, Truscott achieved considerable success. Operation Dragoon shattered the German Nineteenth Army and captured the ports of Marseille and Toulon which became major logistical hubs vital to the success of the Allies in France. Had Truscott commanded more than three infantry divisions and received the vital supplies he needed, an even greater success would have been possible.

At the end of 1944 Truscott returned to Italy as the commander of the US Fifth Army. Bad weather and German counterattacks delayed a significant Allied offensive until April 1945, but when it commenced Fifth Army was unstoppable. It drove the Germans from one defensive position to another until their final surrender. Truscott was correct when he assessed Fifth Army's final campaign in Italy as 'one of the really outstanding campaigns of the war'. That it succeeded so well owed much to the man in charge: Lieutenant General Lucian K. Truscott Junior.

Truscott's war service is also unique in one other aspect. No senior Allied commander took part in more amphibious operations. This commenced with Truscott's observer role in the landing at Dieppe in August 1942. He then took an active part in four more amphibious landings, starting on the shores of French North Africa. This was followed by landings on Sicily, the Italian mainland and southern France. It is somewhat of an irony that a soldier who, after the war, still described himself as a 'cavalryman' and whose sailing experience was having twice been in a small boat, should become the US Army's foremost expert on amphibious operations. But it happened and as Russell F. Weigley wrote in relation to Truscott's role in the Operation Dragoon landing: 'It would have been hard to find a better commander for the amphibious assault. . . . Truscott combined a horse soldier's dash with a wealth of amphibious experience.'

His mistakes

Like any other military commander in history, Truscott was not perfect and did make mistakes. His worst tactical error was to use the Rangers to lead the attack on Cisterna at the end of January 1944. It resulted in the complete destruction of two Ranger battalions, the army formation he had done so much to create. Truscott could also have done more to transform the US 92nd Infantry Division into an effective fighting formation, including finding a commander who valued and respected the African American soldiers he commanded. Truscott's affair with Clare Boothe Luce was also a mistake, not to mention a betrayal of Sarah, his 'Beloved Wife'. Truscott was also a heavy drinker and smoked incessantly. While these did not interfere with his performance as a senior officer, in the long term they impacted seriously on his health.

The qualities of a general

Lucian Truscott Junior had once advised his young son that 'every good commander in a war ... every damn one of them has to have some sonofabitch in him'. What he meant was that military commanders, much like polo players, needed mental and physical toughness. In the military environment this meant taking risks and making decisions knowing that, even if it was the correct decision, men were going to die. If it was the wrong decision, more men would die. Nevertheless, the commander had to make these decisions and not let the terrible consequences affect his judgement. Having some sonofabitch in him also meant that the commander needed to stand up to superiors when they were wrong and not be a yes-man. It meant finding good subordinates and then driving them to achieve the mission. If the subordinate could not follow orders or kept failing to achieve the assigned mission, they should be replaced.

There is little doubt that Truscott had this sonofabitch quality. Being a tough general with some sonofabitch in him should not be confused with being callous or unfeeling. Truscott felt the loss of young American soldiers deeply. His Memorial Day speech at the Sicily-Rome American Cemetery immediately after the war is evidence of this. When his son, Lucian Truscott III, was about to depart for the war in Korea, he asked his father what the Second World War was like. Lucian III wrote that: 'Tears came to his eyes when he described the innumerable dead, on both sides, and how he would never forget how young they all were. Even then I think he was overwhelmed by the sheer numbers of the dead.' It revealed a part of the man that had to be kept hidden while in command of thousands of soldiers.

Overshadowed by others

So, if Lieutenant General Lucian Truscott was a major figure in the Allied war effort, why is he not well known? There are two main reasons Truscott has often been overlooked. First, Truscott hated publicity of any kind. Field Marshal Sir Harold Alexander made an astute assessment when he noted that Truscott was 'shy in a crowd'. Truscott never took the opportunity to promote his successes and was uncomfortable doing so. One of Truscott's biographers, H. Paul Jeffers, illustrates this point with an incident that occurred during Operation Dragoon. An American paratrooper had captured a high-ranking German general and proudly marched him to Truscott's headquarters. The paratrooper said to Truscott: 'Here's a general, sir, all served up on toast.' Jeffers writes that this incident was 'a chance for a picture that Patton or Clark would have relished'. Not so Truscott. 'Send him back to the Seventh Army. I've got other things to do,' Truscott ordered the paratrooper. This incident was repeated on a larger scale in early May 1945. The surrendering German generals had passed through Truscott's Fifth Army headquarters, but he had refused to meet with them. Truscott sent them on to Clark's headquarters without delay.

The other reason Truscott is not well known is that he fought primarily in secondary theatres of war. The Americans never wanted to fight in Italy and once Rome was taken their main efforts were focused on northern France. Italy was then a sideshow and an unwelcome one at that. The invasion of southern France in Operation Dragoon was always overshadowed by the earlier June landing in Normandy. It was part of what Churchill labelled as 'the tyranny of Overlord' in that the Normandy landings tended to eclipse everything else. Operation Dragoon never captured the imagination or publicity as Operation Overlord did, and it was an offensive that the British felt was unnecessary. As Truscott told an interviewer in 1959 in relation to Operation Dragoon: 'When you lead the secondary attack you are always overshadowed.' While being eclipsed by the deeds of others never bothered Truscott, it is unfair. It is certainly time now for his achievements to emerge from the shadows.

There is little doubt that Truscott was a superb tactical commander. Indeed, he has solid claims to be considered as the best American combat commander and general of the Second World War. The final word here is from the astute conclusion of Truscott's first biographer, H. Paul Jeffers, who wrote:

> Yet, no one in the war contributed more on the front lines to defeating Germany than the raspy-voiced ex-schoolteacher, ninety-day wonder, and tight-lipped general who liked meals with flowers on the table and did not give a damn if his name was in newspapers or written large in history books.

Truscott might not have given a damn, but he deserves better.

Epilogue: Taps

Reflecting on his time as a young cavalry officer in charge of the honour guards at the National Cemetery in Arlington, Virginia, Lucian Truscott wrote that:

> A military funeral is a beautiful and impressive ceremony . . . One learns that there is beauty in death and in the respect accorded the dead by the nation and those who live after them. And when the band turns away, leaves the ceremony, and strikes up a sprightly quick step to lead the escort onto other duties, it is a symbol that the living must carry on.

The bugle call Taps that is played at the end of a military funeral, although of recent origin from the American Civil War, was to Truscott the most moving part of a military funeral. He wrote that Taps was 'the most beautiful of all bugle calls' and that: 'There can be no more fitting honour to a departed comrade or to the end of a funeral ceremony than this call which recalls the words:

> Fades the light,
> And afar,
> Goeth light,
> Cometh night,
> And a star,
> Leadeth all,
> Speedeth all,
> To their rest.'

On Wednesday, 15 September 1965, General Lucian Truscott Junior was buried in full dress uniform on a gentle slope a few feet away from the Fort Meyer gate under some large oak trees. A black horse had followed the casket with the boots in the stirrups and the sabre reversed, a cavalry tradition to show that the warrior's fighting days were now at an end. When his body was laid in the earth, seven soldiers, immaculately dressed, fired three volleys in the air and then the bugler played Taps. Sarah Truscott, who would join her husband at his last resting place in 1974, later recalled that this rendition of Taps was 'more beautifully blown than I ever heard it'.

Had he been able to see it, General Truscott would have been immensely pleased with his last parade. It was indeed 'a beautiful and impressive ceremony'. He could not have wanted anything else.

A note on sources for this book

Several centres of research provided the primary source material for this book. In the USA, three institutions, in particular, held a wealth of primary source material relating to Lucian Truscott Junior. Rick Atkinson has described the US Army's Military History Institute, now part of the Army Heritage and Education Centre at Carlisle, Pennsylvania, as 'among the nation's finest archival repositories and *the* mother lode of Army history'. I certainly found this to be true. However, when it came to Lucian Truscott Junior, equally fine and no less fruitful was the vast archive of material held by the George C. Marshall Research Library in Lexington, Virginia. This included the private papers of many senior US officers of the Second World War as well as oral history transcripts, war diaries and other military records, along with documents copied from the US National Archives. The bulk of Lucian Truscott's papers are at the George C. Marshall Research Library. Finally, the papers of Jacob Devers were held by the York County Heritage Trust Library and Archives, York, Pennsylvania.

I am grateful for the assistance, courtesy and professional service provided by the staff of all these research centres.

The material used for this book came primarily from the sources listed below.

The United States Army Heritage and Education Centre (AHEC)
Mark W. Clark Papers
John W. O'Daniel Papers Personal Files
Hamilton Howze Papers
John P. Lucas Papers
Lucian K. Truscott Papers

Ben Harrell Papers
Diary of LT GEN Jacob L. Devers
Senior Officers Oral History Program. Interviews include Mark W. Clark, Ira C. Eaker and Jacob Devers.

George C. Marshall Research Library
George C. Marshall Papers
Lucian K. Truscott Jnr Papers
Field Marshal Harold Alexander, Answers to George Howe's questions for Alexander.
Oral Histories, Marshall Library. The Oral Histories interviews include George C. Marshall, Field Marshall Lord Wilson, Mark W. Clark, Jacob L. Devers and Lucian K. Truscott.

York County Heritage Trust Library and Archives
Papers of Jacob Devers

In relation to the secondary sources consulted, the three existing biographies of Truscott were invaluable. For a thorough analysis and understanding of the Italian campaign, the two books by James Holland were very useful while Rick Atkinson's *The Day of Battle* remains in a class of its own. I have also extensively used Truscott's two publications *Command Missions* and *The Twilight of the U.S. Cavalry*, the latter being particularly useful for Truscott's army career up to 1942. The Editor's Preface of *The Twilight of the U.S. Cavalry*, written by Lucian Truscott III, also provided a unique insight into his parents' character and their marriage.

Further reading/watching: Lieutenant General Lucian Truscott Jr

Rick Atkinson, *The Day of Battle: The War in Sicily and Italy, 1943–1944*, New York, Henry Holt and Company, LLC, 2008.

Gregory Blaxland, *Alexander's Generals: The Italian Campaign 1944–45*, London, William Kimber, 1979.

Robert M. Citino, *The Wehrmacht's Last Stand: The German Campaigns of 1944–1945*, Lawrence, University Press of Kansas, 2017.

Carlo D'este, *Fatal Decision*, London, HarperCollins Publishers, 1991.

Harvey Ferguson, *The Last Cavalryman: The Life of General Lucian K. Truscott, JR.*, Norman, University of Oklahoma Press, 2015.

Wilson A. Heefner, *Dogface Soldier: The Life of General Lucian K. Truscott, Jr.*, Columbia, University of Missouri Press, 2010.

James Holland, *Italy's Sorrow: A Year of War, 1944–45*, London, HarperPress, 2008.

James Holland, *The Savage Storm: The Battle for Italy 1943*, London, Penguin Random House, 2023.

H. Paul Jeffers, *Command of Honor: General Lucian Truscott's Path to Victory in World War II*, New York, New American Library, 2008.

L.K. Truscott, Jr., *Command Missions: A Personal Story*, New York, E.P. Dutton & Company, Inc., 1954.

Lucian K. Truscott Jr, *The Twilight of the U.S. Cavalry: Life in the Old Army, 1917–1942*, Lawrence, University of Kansas Press, 1989.

Russell F. Weigley, *Eisenhower's Lieutenants: The Campaigns of France and Germany, 1944–1945*, London, Sidgwick & Jackson, 1981.

YouTube, https://www.youtube.com/watch?v=_HlJxM54Ex0 *The Greatest Field Commander: Lucian Truscott*

Photographic credits

Page 21: *Hot action in the polo match today between the teams of the 6th Field Artillery and the 16th Field Artillery. This was the first match of the War Dept. Polo Association Tournament opening today.* 1926. Photograph. https://www.loc.gov/item/89711913/. United States Library of Congress.

Page 27: Harris & Ewing, photographer. *TRUSCOTT, LUCIAN K. CAPTAIN*, None. [Between 1905 and 1945] Photograph. https://www.loc.gov/item/2016862652/, United States Library of Congress.

Page 52: Maj-Gen L.K. Truscott, Commanding General 3rd Division, Box 282, Folder Truscott, Lucien K., Personality Collection, U.S. Army Heritage and Education Center, Carlisle, PA. U.S.

Page 53: *Allied leaders inspect invasion forces / Official U.S. Navy photograph.* Sicily Italy, 1943. Photograph. https://www.loc.gov/item/92512686/. United States Library of Congress.

Page 55: *Advance inland--through the narrow streets of Scoglitti, the powerful spearhead of allied invasion forces--in this instance American troops--begin their triumphal surge over the western and northern sector of Sicily.* Sicily Italy, 1943. Aug. 2. Photograph. https://www.loc.gov/item/89712757/. United States Library of Congress.

Page 57: *George Patton.* 1943. Photograph. https://www.loc.gov/item/2004672054/. United States Library of Congress.

Page 72: A general conference in the sleeping hours of morning, Box 282, Folder Truscott, Lucien K., Personality Collection, U.S. Army Heritage and Education Center, Carlisle, PA. U.S.

Page 85: John W. O'Daniel Papers, Personal Files, Box 1, U.S. Army Heritage and Education Center, Carlisle, PA. U.S.

Page 97: Maj-Gen Lucian Truscott CG of VI Corps confers with staff officers, Box 282, Folder Truscott, Lucien K., Personality Collection, U.S. Army Heritage and Education Center, Carlisle, PA. U.S.

Page 98: Maj-Gen Lucian Truscott CG of VI Corps and Ernie Pyle, Box 282, Folder Truscott, Lucien K., Personality Collection, U.S. Army Heritage and Education Center, Carlisle, PA. U.S.

Page 107: Maj-Gen Lucian Truscott CG of VI Corps, Box 282, Folder Truscott, Lucien K., Personality Collection, U.S. Army Heritage and Education Center, Carlisle, PA. U.S.

Page 110: Maj-Gen Lucian Truscott at VI Corps HQ, Box 282, Folder Truscott, Lucien K., Personality Collection, U.S. Army Heritage and Education Center, Carlisle, PA. U.S.

Page 133: [PHOTO] LT GEN LUCIAN K TRUSCOTT, VI CORPS COMMANDER; LT GEN ALEXANDER M PATCH, CG, 7TH ARMY; AND LT GEN JACOB L DEVERS, CG, 6TH ARMY GROUP, TALK AT THIRD DIC CP. SEVENTH ARMY, FRANCE Box 1B, Folder 20, William W. Quinn Collection, U.S. Army Heritage and Education Center, Carlisle, PA. U.S. https://emu.usahec.org/alma/multimedia/1684803/20184043PHBW1013041450F000000033 0795I 013.pdf)

Page 139: Lt-Gen Lucian Truscott at his desk, Box 282, Folder Truscott, Lucien K., Personality Collection, U.S. Army Heritage and Education Center, Carlisle, PA. U.S.

Page 141: WIDE SHOT OF LIEUTENANT GENERAL LUCIAN K. TRUSCOTT; MAJOR GENERAL GEOFFREY KEYES WALKING DOWN ROCKY HILL WITH TENTS BEHIND THEM. Box 1, Folder 15, Geoffrey Keyes Photograph Collection, U.S. Army Heritage and Education Center, Carlisle, PA. U.S. https://emu.usahec.org/alma/multimedia/520813/20183171PHBT1036357 602F364179I012.pdf

Page 143: Lt-Gen Lucian Truscott and soldiers of the 92nd Division, Box 282, Folder Truscott, Lucien K., Personality Collection, U.S. Army Heritage and Education Center, Carlisle, PA. U.S.

Page 144: MEDIUM SHOT PHOTOGRAPH OF MAJOR GENERAL

EDWARD ALMOND RECEIVING AWARD FROM LIEUTENANT GENERAL LUCIAN TRUSCOTT. Box 3A, Folder 19, Edward M. Almond Photograph Collection, U.S. Army Heritage and Education Center, Carlisle, PA. U.S. https://emu.usahec.org/alma/multimedia/1776520/20181428PHBT9891 08386F0000000203629I005.pdf

Page 149: WIDE SHOT OF LIEUTENANT GENERAL LUCIAN K. TRUSCOTT; LIEUTENANT GENERAL GEOFFREY KEYES SHAKING HANDS WHILE STANDING ON GRASS FIELD WITH SOLDIERS STANDING AROUND THEM, Box 1, Folder 6, Geoffrey Keyes Photograph Collection, U.S. Army Heritage and Education Center, Carlisle, PA. U.S. https://emu.usahec.org/alma/multimedia/520577/20183171PHBT1036357602F364143I002.pdf

Page 151: Van Vechten, Carl, photographer. *Portrait of Clare Boothe Luce*, 1932. Dec. 9. Photograph. https://www.loc.gov/item/2004663224/. United States Library of Congress.

Page 166: Lt-Gen Lucian Truscott and soldiers Fifth Army, Box 282, Folder Truscott, Lucien K., Personality Collection, U.S. Army Heritage and Education Center, Carlisle, PA. U.S.

Page 173: Lt-Gen Lucian Truscott, Box 282, Folder Truscott, Lucien K., Personality Collection, U.S. Army Heritage and Education Center, Carlisle, PA. U.S.

Acknowledgements

There are many individuals and organizations who help bring a book such as this to publication. This book originated with an idea of Gareth St John Thomas of Exisle Publishing to create a series of books on lesser-known battlefield commanders that analysed the essential elements of their command style. Each volume was to be a succinct study that analysed the commander's military leadership and their success on the battlefield; one that could be read in one or two sittings. I am immensely grateful that Gareth invited me to write the first two volumes of this series. Gareth has been a strong supporter of this volume on Lucian Truscott Junior, and I thank him for his continued interest and enthusiasm.

Exisle Publishing have been a pleasure to work with. I especially want to thank Anouska Jones, the publisher and project manager, for her dedication, skill and professionalism throughout the publishing process. Brian O'Flaherty was the copy editor for the manuscript and made the process as smooth as possible. Nick Turzynski was responsible for the book design and produced the excellent maps.

The primary research for this book was undertaken at the George Marshall Research Library in Lexington, Virginia, the US Army Heritage and Education Centre (US AHEC) at Carlisle, Pennsylvania, the US Library of Congress, Washington DC, and the York Historical Society Museum, York Pennsylvania. Without fail, the staffs of these institutions were friendly, courteous and helpful.

My friend Professor Michael S. Neiberg of the US Army War College helped track down research material in the United States. It was Michael who put me in touch with Justine Melone of the Collection Directorate of the US AHEC. Justine tracked down most of the photographs used in the book.

Finally, I wish to acknowledge and thank my wife, Susan Lemish, for her support and for the work she has put into this book. Susan read all the first drafts and wielded her red pen for corrections with gusto. The book is much better for her input. Susan also assisted with compiling the index which most authors will acknowledge is the least enjoyable part of the writing process. I could not have completed the book on time without Susan's assistance.

Glyn Harper
Foxton Beach
New Zealand
January 2025

Index

Abrams, William R. 166
African-Americans 142–7, 158, 183
Agrigento 58
Aix-en-Provence 129
Alban Hills (Colli Laziali) 75, 76, 81, 82, 83, 86, 95–103, 104, 108, 109, 112, 113
Albano 81, 83, 86, 104
Alexander, Harold R.L.G. 'Alex' 53, 58, 59, 69, 75, 82, 84, 85, 91, 94, 95, 96, 100, 102, 104, 112, 138, 140, 142, 180, 184
Algeria 51
Algiers 46
Allied Military Formations:
  Sixth Army Group 126, 133, 135
  Twelfth Army Group 135
  Fifteenth Army Group 138, 159, 164
  Twenty-first Army Group 159
  Eastern Military District (Bavaria) 169, 170
  Supreme Headquarters Allied Expeditionary Force (SHAEF) 135, 136, 138, 140
Almond, Edward M. 144, 146
Antwerp 136
Anzio 9–12, 70, 74–118, 119, 140, 148, 165, 181
Apennine Mountains 66, 147, 148, 156, 159, 160, 162, 163
Aprila (The Factory) 81, 83, 88, 93
Arizona 16, 17, 20, 119
Arlington, Virginia 26
Arlington National Cemetery 26, 186
Arkansas 15, 16, 32
Army War College 26
Aspres-sur-Buéch 129
Astura River 81
Atkinson, Rick 61, 68, 82, 94, 104, 107, 131, 146, 180
Avignon 123

Baehr, Carl 93, 97, 100
Barga 153, 154
Bavaria 169, 170
Belfort 131, 132, 134
Belgium 22, 139, 153, 154, 172
Belvedere Ridge 157
Berlin 114, 126
Besançon 131, 132
Bluemont 174
Bologna 147, 148, 154, 156, 159, 160, 162, 163
Bonaparte, Napoleon 146
Bondeno 162, 163
Bonus Army 28
Bradley, Omar Nelson 54, 56, 59, 60, 62, 134, 135
Bradshaw, Aaron 96

Brann, Donald 110, 111
Brazilian Forces:
  1st Brazilian Infantry Division 142, 160, 161
Brignoles 129
British Military Units/Formations:
  Armies:
    First Army 46
    Eighth Army 56, 59, 65, 102, 103, 115, 142, 147, 148, 159, 160, 162, 163, 164
  Corps:
    XIII Corps 142
  Divisions:
    1st Division 76, 79, 83, 85–6, 103, 105
    5th Division 99, 105
    8th Indian Division 153–4
    56th London Division 89, 99, 103
  Brigades:
    2nd Special Services Brigade 79, 118
    Combined Operations Headquarters 35, 36, 37, 39
    Commandos 35, 37, 38, 49, 79
    HMS *Fernie* 40
Brolo 60
Brooke, Alan 102
Bruyéres 134
Butler, Fred B. 123, 129

Caeser Line 112, 113, 116
California 32, 150
Camp Hale 148
Camp Harry J. Jones 17
Camp Marfa 23
Campoleone 84, 85, 86, 88, 104
Canadian Corps 159
Canaris, Wilhelm F. 80
Cape Nègre 122, 124
Capri 154
Carrara 160
Casablanca 40
Caserta 164
Cassino 75, 102, 112, 147, 164
Castelnuovo di Garfagnana 153
Cavalry School 23, 25
Central Intelligence Agency (CIA) 174–5, 177
Charlottesville, Virginia 18, 37
Chatfield, Texas 13
China 168, 175
Chou En-Lai 175
Churchill, Randolph 150
Churchill, Winston 75, 77, 81, 85, 94, 112, 121, 137, 140, 185

Cisterna 81, 83, 84, 85, 86, 87, 88, 100, 104, 105, 106, 108, 109, 183
Citino, Rob 42, 75, 94, 136, 180
Clark, Mark W. 9, 10, 33, 35, 36, 39, 65, 72, 73, 75, 77, 78, 82, 84, 85, 87, 88, 90, 91, 94, 95, 99, 100, 108, 111, 112, 114, 115, 118, 120, 138, 140, 141, 145, 146, 149, 153, 155, 158, 159, 163, 164, 180, 184
Colorado 119
Cold War 175
Columbus, Oklahoma 15
Combined Chiefs of Staff Committee 138
Command Missions 20, 35, 171, 175, 181
Command and General School 29–31
Como 163
Connecticut 150
Cori 81, 109
Costello, John M. 150
Crosby, Herbert B. 25
Crittenberger, Willis 140, 142, 146, 149, 150
Czechoslovakia 170

Dachau 170, 171–2
Dahl, Roald 151, 152
Dahlquist, John E. 120, 126, 129, 130
Daniel, John L. 166
Darby, William Orlando 38, 39, 80, 82, 86, 87
D'Este, Carlo 180
de Lattre de Tassigny, Jean 123, 126
Devers, Jacob L. 95, 96, 117, 120, 126, 133, 135, 145, 174, 180
Dieppe 38, 39–40, 42, 182
Dijon 122, 136
Dill, John 138
Doolittle, James H. 'Jimmy' 54
Doubs River 131
Douglas, Arizona 16, 17, 23
Dulles, Allen 174

Eagles, William 92, 93, 119
Eastland, William H. 166
Eisenhower, Dwight D. 'Ike' 26, 28, 32, 35, 36, 39, 46, 53, 94, 95, 121, 126, 134, 135, 139, 140, 168, 169, 175, 176, 179, 181
Épinal 134
Eufaula, Oklahoma 14
Everleigh, Vivian 96

Ferguson, Harvey 28, 77, 152
Fidenza 164
First World War 15, 16, 48, 70, 145
Florence 153, 154, 164
Fort Bliss 33
Fort Knox 26, 31, 32
Fort Leavenworth 29, 31
Fort Lewis 32, 33
Fort Logan H. Roots 15
Fort Myer 25, 26, 28, 187
Fort Riley 23, 24

Fort Washington 28
Fortune Magazine 150
France 17, 40, 54, 83, 116, 117, 119, 121, 134, 137, 138, 140, 153, 182, 185
Frankfurt 168, 174, 175
Fredenhall, Lloyd 40, 46, 47
Fréjus 123, 124
French Military Forces:
    First Army 126
    French Expeditionary Corps (FEC) 103, 119, 148
    II Corps 121, 122, 126
    1st French Armoured Division 123

Gallipoli 9, 11, 76
Gap 129
Garfagnana, Battle of 153
Genoa 160
Georgia 18
German Military Units/Formations:
    Armies:
        Tenth Army 68, 103, 104, 111, 112, 116
        Nineteenth Army 123, 126, 136, 137, 182
    Corps:
        1st Parachute Corps 111, 113
    Divisions:
        3rd Panzer Grenadier Division 112
        4th Parachute Division 112
        11th Panzer Division 127–9
        26th Panzer Division 89
        29th Panzer Grenadiers 60, 89
        65th Infantry Divion 112
        Hermann Göring Panzer Division 83, 84
    Regiments:
        Lehr Regiment 90
Germany 15, 22, 83, 139
Grenoble 127, 129
Gruenther, Alfred M. 78
Gothic Line 147, 148, 153, 154–5, 159, 160, 162
Guadalcanal 120
Gustav Line 75, 102, 104, 108, 147

Halifax, Lord (Wood, Earl) 152
Hamm 170
Harding, A.F. 'John' 94, 102
Harmon, Ernest N. 'Ernie' 86, 92, 93, 103, 106, 108, 109
Hawaii 18, 20, 22, 23
Hays, George P. 156
Heefner, Wilson A. 146
Heidelberg 174
Hewitt, Kent 41
Highway 6 (Via Casilina) 66, 81, 86, 103, 104, 108, 109, 110, 111, 114
Highway 7 (The Appian Way) 66, 81, 83, 86, 104, 105, 106, 113
Highway 9 162, 163
Highway 64 156, 162
Highway 65 162
Hitler, Adolf 83, 88, 89, 94, 99, 136

Hodges, Courtney 72
Hornbrook, James H. 17
Hoover, Herbert 28
Howze, Hamilton H. 109, 111, 180

Italian Forces
    Italian Combat Group Legnano 142, 162
Italy 65, 66, 70, 76, 80, 116, 118, 120, 121, 124, 134, 137, 138, 140, 142, 145, 146, 148, 154, 155, 160, 165, 168, 170, 181, 182, 185

Japan 168
Jeffers, H. Paul 50, 184, 185
Jefferson, Thomas 18
Juin, Alphonse Pierre 119
Jura Mountains 132

Kansas 29
Kasserine Pass 46
Kennedy, Joseph P. 150
Kentucky 26, 31
Kerwin, Walter 'Dutch' 93
Kesselring, Albert 80, 82-3, 85, 88, 89, 90, 94, 99, 100, 113
Keyes, Geoffrey 60, 108, 138, 139, 141, 142, 149, 150, 168, 172
Kirk, Alan G. 53
Korean War 184

Ladue, Laurence 99
Lake Garda 163
Lang, Will 135
Larma, Rosario V. 166
La Spezia 159, 160
Legnano 163
Le Muy 122, 127
Licata 54, 56
Life Magazine 16, 135, 150
Little Rock, Arkansas 15
Littoria 81
Livorno (Leghorn) 153
Louisiana 32, 33
Lucas, John Porter 9, 10, 11, 47, 68, 69, 70, 73, 76, 77, 78, 81, 82, 83, 84, 85, 87, 88, 90, 91, 92, 94, 95
Luce, Clare Boothe 150-2, 183
Luce, Henry Robinson 150, 152
Luxembourg 135, 153, 170
Lyon 122, 131

MacArthur, Douglas 2 5, 26, 28
Mackensen, Eberhard von 88, 89, 90, 99
Malmedy 172
Marrakech 77
Marseille 121, 122, 123, 126, 129, 136, 182
Marshall, George C. 34, 35, 36, 38, 50, 94, 95, 138, 145, 149, 154-5, 168
Massa 159, 160
Mauldin, Bill 124, 167, 179

McCreery, Richard L. 164
Mehdia 45
Messina 58, 59, 60, 63, 64
Mexico 15, 16
Mignano Gap 70, 73
Modena 162
Moletta River 81, 82
Monte Caruso 68
Monte Cassino 73
Monte Cipolla 60, 62, 64
Monte La Difensa 70
Montélimar 127, 129, 130, 131
Monte San Michele 162
Monterey, California 23
Montese 161
Montgomery, Bernard Law 56, 59, 60, 63, 65, 134, 159
Monticello 18
Morrison, Samuel E. 76
Mount Belvedere 156
Mount Cesima 73
Mount Lungo 70, 73
Mount Rotondo 73
Mountbatten, Louis 35, 36, 37
Munich 172
Murphy, Audie 48, 50
Mussolini Canal 86, 105

Naples 65, 66, 68, 69, 77, 78, 81, 101, 122, 148
National Defense Act (1916) 15, 16
National Defense Act (1920) 22
Nettuno 10, 79, 80, 82, 83, 95, 97, 102, 104, 165
New Caledonia 120
New Mexico 15, 119
Nisei soldiers 147
Norman, Oklahoma 14
Normandy 121, 122, 185
North Africa 38, 40, 42, 43, 45, 49, 179, 181, 182
North Carolina 48
Nuremburg trials 170-1

O'Daniel, John W. 'Iron Mike' 51, 90, 91, 100, 119, 131
Oklahoma 13, 14
Onapa, Oklahoma 14
Operation Anvil 121
Operation Buffalo 104, 105, 109, 110
Operation Diadem 102, 103, 104, 112
Operation Dragoon 119-26, 135, 136, 182, 184, 185
Operation Fischfang 88-90, 93, 94, 99
Operation Fourth Term 145
Operation Goalpost 40-1, 45
Operation Grapeshot 159, 162
Operation Husky 51-68, 76
Operation Jubilee 39
Operation Overlord 54, 95, 121, 136, 185
Operation Shingle 75, 76, 77, 78, 79-82
Operation Torch 38, 40-7, 181

INDEX | **199**

Operation Turtle 104, 111
Operation *Wintergewitter* (Operation Winterstorm) 153, 154, 155

Padiglione Woods 89
Padua 163
Palermo 56, 58, 65
Paris 48
Parma 164
Patch Alexander M. 'Sandy' 120, 121, 122, 123, 124, 127, 132, 133, 134, 135
Patton, George S. Jr. 22, 26, 31, 36, 38, 39, 40, 41, 42, 45, 46, 47, 53, 54, 56, 57, 58, 59, 60, 61-3, 64, 130, 131, 135, 168-70, 179, 184
Perks, Troy J. 166
Pearl Harbor 33
Penney, Ronald 76
Piacenza 164
Po River 159, 162, 163
Po Valley 159, 163
Ponte Samoggia 162
Pontine Marshes 81
Port Lyautey 40, 41, 42, 45, 48, 53, 126, 181
Porto Empedocle 58
Pyle, Ernie 69, 98

Randolph, William Mann 18
Regio 65
Rhine River 132
Rhône Valley 122, 123, 126, 127-131
Rhône River 127, 132
Riva Ridge 156, 157
Rocco di Roffeno 161
Rocco Pipirozzi 70
Rome 9, 66, 75, 76, 79, 81, 83, 87, 95, 103, 104, 108, 109, 111, 112, 114, 115, 116, 121, 147, 154, 181, 185
Roosevelt, Franklin D. 29, 134, 140
Ryder, Charles W. 'Doc' 40

Saint Tropez 122
Sainte Maxime 122
Sale 45
Salerno 65, 75, 76
Samoggia River 162
San Benedetto 163
San Francisco, California 17
San Fratello 60
San Stefano 60
Scoglitti 55, 56
Sebou River 41
Second World War 23, 30, 31, 32, 147, 179, 184, 185
Senger und Etterlin, Fridolin von 164
Serchio River 154
Serchio Valley 153, 154, 155, 156
Sicily 19, 50, 51-64, 76, 130, 170, 181, 182
Silk, George 135
Sisteron 127, 129

Smith, Walter Bedell 135, 140, 168, 174
Sorrento 154
South African Forces:
  6th South African Armoured Division 142, 162, 163
Stella, Oklahoma 14
Sturrock, Donald 152

Taps 186-7
Taranto 65
Texas 119
The Dogface Soldier 49-50, 79
The Truscott trot 49, 59
The Twilight of the U.S. Cavalry: Life in the Old Army, 1917-1942 175, 176
Tiber River 115
Time Magazine 150
To Hell and Back 48, 50
Tomb of the Unknown Soldier 26
Toulon 121, 122, 123, 126, 136, 182
Truman, Harry S. 174
Truscott, Jaimie 25, 37
Truscott, Lucian K. Snr. 13, 14
Truscott, Lucian K. Jnr,
  Anzio 9-12, 79-82, 92-115
  CIA role 174-5, 177
  Clark, Mark, relations with 87-8, 98-9, 117
  Death of 178, 186-7
  Dieppe 39-40
  Drinking 10, 64, 101, 172, 177, 183
  Early life 13-29
  Eisenhower, relations with 26, 28, 32, 35, 36, 39, 46, 168, 175, 181
  Fifth Army, command of 139, 142-165
  Fifteenth Army, command of 134, 136, 137
  France (1944 campaign) 124-34, 136-37
  Health 9, 82, 84, 101, 172-5, 177-8, 183
  Italy campaign 11, 65-116, 140-167
  Luce, Clare Boothe 150-2, 183
  Marriage/family 18-20, 24-5, 176-7, 183
  Marshall, George, relations with 34, 35, 36, 38, 138, 154-5
  Memorial Day Speech (1945) 165-7, 184
  Military postings (USA) 22-31
  North Africa 41-7
  Operation Dragoon 121-6
  Operation Husky 51-68
  Operation Goalpost 40-1, 45
  Operation Shingle 79-82
  Operation Torch 40-7
  Patton, George, relations with 61-3, 168-70
  Polo 20-2, 29, 31, 36, 183
  Promotions 15, 20, 22, 30, 31, 33, 37, 46, 133, 134-5, 176
  Retirement 176-8
  Sicily invasion 51-64
  War Department duties 174
Truscott, Lucian K. III (Lucian III) 16, 18, 19, 20, 21, 24, 31, 117, 176, 177, 184

Truscott, Lucian K. IV (Lucian IV) 19, 135
Truscott, Maria 13, 14, 23
Truscott, Mary 177
Truscott, Patsy 14
Truscott, Sarah 10, 18-20, 24, 26, 31, 37, 40, 74, 102, 109, 114, 116, 135, 152, 172, 174, 176, 177, 183
Truscott, Trott 49, 59
Tuchman, Barbara 11
Tunisia 46

United Nations 172
United States Units/Formations:
  Armies:
    Third Army 131, 168, 169, 170, 172
    Fifth Army 9, 65, 66, 68, 70, 77, 99, 102, 103, 104, 107, 108, 109, 112, 115, 138, 139, 140, 142, 145, 147, 148, 149, 150, 151, 154, 155, 156, 157, 159, 160, 162, 163, 164, 165, 168, 182, 184
    Seventh Army 54, 56, 58, 59, 60, 64, 118, 119, 121, 126, 133, 135, 184
    Fifteenth Army 134, 136, 137, 142, 170
  Corps:
    II Corps 46, 54, 59, 108, 114, 142, 160, 162
    IV Corps 142, 160, 161, 162, 163
    VI Corps 9, 10, 11, 65, 66, 68, 76, 77, 92, 94, 95, 97, 103, 104, 105, 108, 109, 110, 111, 112, 113, 114, 115, 116, 117, 119, 121, 123, 124, 126, 130, 131, 132, 134, 135, 136, 142, 148, 181, 182
  Divisions:
    1st Armored Division 31, 38, 76, 84, 86, 92, 93, 103, 105, 106, 109, 111, 112, 115, 142, 153, 160, 161, 162, 163
    1st Cavalry Division 33
    1st Infantry Division 59, 65
    2nd Armored Division 54
    3rd Infantry Division 9, 11, 47, 48-51, 52, 54, 58, 59, 60, 64, 65, 66, 68, 69, 70, 73, 74, 76, 77, 78, 79, 80, 83, 86, 87, 88, 90, 91, 92, 93, 96, 100, 103, 105, 106, 107, 108, 109, 110, 111, 112, 114, 118, 122, 126, 129, 130, 131, 181
    9th Infantry Division 65
    10th Mountain Division 142, 148, 155-8, 160, 162, 163
    34th Infantry Division 38, 99, 103, 105, 111, 142, 160, 162, 163, 166
    36th Texas Division 73, 103, 105, 111, 112-4, 115, 118, 122, 124, 126, 127, 129, 130, 131
    45th Infantry Division 59, 76, 84, 88, 89, 90, 92, 93, 103, 105, 106, 111, 118, 122, 126, 129
    85th Infantry Division 114, 115, 142, 153, 160, 162, 166
    88th Division 142, 160, 163, 166
    91st Infantry Division 142, 160, 163, 166
    92nd Infantry 'Buffalo' Division 142-7, 153, 154, 155, 158, 159, 160, 166, 183
  Brigades:

1st Cavalry Brigade 26

Regiments:
  3rd Cavalry Regiment 26
  5th Cavalry Regiment 33
  7th Infantry Regiment 58, 86
  11th Cavalry Regiment 23
  13th Armored Regiment 31
  15th Infantry Regiment 58, 86
  17th Cavalry Regiment 16, 17, 22
  30th Infantry Regiment 73
  60th Infantry Regiment 41, 43
  85th Mountain Regiment 158, 163
  86th Mountain Regiment 157
  442nd Regiment 147
  473rd Armored Regiment 147
  504th Parachute Regiment 86
Battalions:
  1st Ranger Battalion 39, 82
  3rd Ranger Battalion 82
  4th Ranger Battalion 82
  509th Parachute Infantry Battalion 82
Other units:
  1st Special Force 84, 103, 105, 106, 110, 111, 112, 122
  8th Bomber Command 53
  56 Evacuation Hospital 102
  Butler Force 123, 127, 129
  Joss Force 54, 56
  Rangers 39, 49, 54, 58, 80, 183
  The Invasion Training Center 51
  Western Task Force 40, 42, 46
  USS *Biscayne* 79
  USS *Catoctin* 124

Valmontone 81, 86, 103, 104, 108, 109, 111, 112, 114, 116
Velletri 108, 113
Venice 163
Vergato 161
Verona 163
Versailles 135
Vicenza 163
Villa, Pancho 15
Virginia 18, 26
Virginia Military Institute 22
Volturno River 66-70
Vosges Mountains 132, 134

Walker, Fred L. 112, 113, 120
Walter Reed General Hospital 172, 174, 178
War Department 34, 35, 174
Washington (state) 32, 33
Washington D.C. 33, 34, 35, 172
Watson, Leroy H. 170
Weigley, Russell F. 30, 54, 120, 137, 182
West Point Military Academy 14, 15, 19, 22, 31, 38
Westphal, Siegfried 100
World War Veterans Act (Bonus Bill) 26